Marconi at The Lizard

The story of communication systems at Housel Bay

Marconi at
The Lizard

The story of communication systems at Housel Bay

by Courtney Rowe, MBE BSc

On 23rd January 1901 Guglielmo Marconi, in his quest to extend the range of wireless telegraphy over the horizon, succeeded in receiving a signal at The Lizard Wireless Telegraphy Station from his station on the Isle of Wight some 300km away. This was a very significant increase over anything that had been done before in a long distance operation and was, for the first time, far over the horizon; it achieved what scientists of the day had considered to be impossible. The publishing of this book marks the centenary of the event.

The Lizard Wireless Telegraphy Station located at Housel bay, The Lizard was an operating ship-to-shore station earning money for the young Marconi Company; in addition it undertook important development work. Other communication facilities have opened at Housel Bay, before and after the Marconi station, and this book surveys the history of these as well as that of the Marconi station.

The Trevithick Society

First Printed 2000
Reprinted 2002, 2009
Copyright © Courtney Rowe

ISBN 0 904040 80 7

Published by the Trevithick Society

Printed by R. Booth Ltd.
The Praze, Penryn, Cornwall TR10 8AA

Typeset by Peninsula Projects
29 Tolver Road, Penzance TR18 2AQ

Contents

List of Figures and Plates

Figures

Plates

The Lizard Wireless Telegraphy Station from the sea and an internal view, together with an announcement by the Marconi Company of February 1901 of the first over-the-horizon wireless telegraphy.

MARCONI'S WIRELESS TELEGRAPH
COMPANY, LTD
TELEGRAMS.
EXPANSE, LONDON.
A.B.C. CODE USED.
TELEPHONE Nº 2748 AVENUE.

18, Finch Lane,
London, E.C.

11th February, 1901.

Dear Sir (or Madam),

I am instructed to inform you that
Mr. Marconi has established communication
between St. Catherine's (Isle of Wight)
and the Lizard (Cornwall) a distance of
196 miles. It is a matter of congratula-
tion that this long distance has been
successfully accomplished.

I am,

Yours faithfully,

HENRY W. ALLEN,

Assistant Manager & Secretary.

All three items by kind permission of Marconi plc

viii

Introduction

The Lizard

Whilst The Lizard has kept much of its old character, there are two particular things that have changed in recent times. The first of these relates to the extent of the area known as The Lizard, which is a small area some 1.6 km (1 mile) square coinciding with the parish of Landewednack, with The Lizard Village at its centre and with a coastline of some 5 km (3 miles) including Housel Bay. The area to the north gradually became known as The Lizard district or peninsula, and now villages that are miles away like to say they are in The Lizard.

The second change is on the prefix "The" on the name The Lizard which locals (and Trinity House, Marconi, the Post Office and the Royal National Lifeboat Institution) have used for many years even though maps say just Lizard. However, those who prepare road and bus signs and holiday guides have recently dropped the prefix, and whilst the Post Office stamp still says The Lizard it is no longer used on letters. The author trusts that the reader will not mind if he keeps to the old ways (which are still the modern ways of the locals) and uses "The Lizard" and with its local definition of extent.

Bay of Communications

The Lizard, the most southerly point of the British Isles was, until the end of the nineteenth century, a remote part of the country, yet its coast was adjacent to one of the busiest shipping lanes with The Lizard Head a point of landfall for ships which were trading throughout the world. The rugged cliffs although picturesque were a great danger to these ships at night and in stormy weather or when they were shrouded in mists, and many ships have foundered on them. The local people earned their living principally in farming, fishing and associated trades. In the better weather the farmers would be in the fields with their horses whilst out at sea the local fishermen would be in their sail boats working against the background of sail and steam trading and naval vessels passing by.

The arrival of the railway at Helston in 1887 and the introduction of motorised bus services in 1903 led gradually to an increased number of visitors to the area. In more recent years the numbers have been swelled by the convenience of the motor car such that The Lizard is now a popular holiday spot visited by about half a million people each year, mostly in the summer months.

Immediately to the east of The Lizard Head is Housel Bay, a mere 1 km (0.6 mile) across, which over the years and particularly in the nineteenth and twentieth centuries attracted a range of developing communication systems several of which were to provide assistance to passing ships. Because of the wide variety in these systems the bay could be called the "Bay of Communications". A map of the bay and its vicinity is shown in Figure 1 and an aerial view in Plate 1. A cliff path which many people now walk goes along the length of Housel Bay and extends beyond it in each direction as far as one pleases.

Many of the passing ships would have been at sea for months and out of sight of land for a long time, and as they approached they would desperately want to avoid an impact with the coast. Yet The Lizard coast at night in the nineteenth century would have been bereft of lights, for the inhabited parts of the village were away from the coast, and mariners approached the shores with trepidation during the hours of darkness. A lighthouse to warn of the presence of rocks and cliffs was set-up at the easterly end of the bay for a brief period in 1619 and became permanent in 1752. This is one of the most well known lighthouses which warns of the nearness and dangers of the coast whilst providing a welcoming sight to those returning from abroad and a point of farewell to those leaving.

But the lighthouse was only effective in fine conditions, many thousands of tons of shipping being lost on the local rocks in foggy weather. In 1878 an audible foghorn, driven by a hot-air engine, was installed at the lighthouse as a first attempt to warn ships in fog of the presence of the coast. However, shipwrecks in fog still remained a major problem and in 1910 Trinity House undertook a further installation by sinking a bell about 2 miles south of The Lizard Head. During fog, this was rung electrically from the lighthouse to guide ships away from the coast; this, in its turn, did not have the desired effect and shipwrecks were to continue for many years when there was fog until ships carried much improved navigation systems.

There was no recognised way of communicating between ships and the shore until 1872 when a shipping agent, G C Fox and Company of Falmouth, built The Lizard Signal Station at Bass Point at the easterly end of the bay. This was used for reporting the passage of ships, and for conveying messages between ships and the station using flags by day and lights and pyrotechnics by night. An overland electric telegraph was installed in the same year, this kept the station in almost instant contact with offices at Falmouth, and elsewhere, and thus information and orders could be conveyed between offices and ships. This station was later taken over by Lloyd's and the building is still there having been restored by the National Trust.

A major step forward was made in 1900 when Marconi constructed a wireless telegraphy station at the bay to give, for the first time, an all-weather and day and night ship to shore communication capability for ships fitted with suitable equipment. This station was also used by Marconi for development work and in January 1901 it achieved wireless telegraphy communication with another station at the Isle of Wight 300 km (186 miles) away, thereby setting-up a new long-distance record and demonstrating for the first time the possibility of "over-the-horizon" wireless propagation which scientists at the time had considered to be impossible. The station also undertook important work in establishing multiple-station operation. The buildings of this station may be seen today having been restored in 2000 by the National Trust.

The arrival of the electric telegraph at The Lizard Signal Station, which is mentioned above, was at a time when electric telegraphy was being extended to countries overseas by using cables under the sea, and attracted the attention of a company which was about to lay a submarine cable between Spain and Cornwall. As a result this cable was brought ashore at Housel Bay in 1872 and terminated in the signal station which became a point of international communication.

During the 1941-1945 period the area between Pen Olver and Bass Point was the location of a Royal Air Force radar station for providing warning of low-flying hostile aircraft and for tracking enemy and friendly ships. The station was known as RAF Pen Olver. A short distance inland on Lloyd's Road is the location of a Royal Observer Corps station used originally between 1941 and 1945 for observing aircraft and reporting their tracks; it was updated to provide service in the Cold-War era and the remains of surface and underground buildings dating from 1951 are still visible.

Each of these communication systems and others are described in the following chapters in the order in which they arrived at the bay. Some are still in operation, while the remains of others can be examined. Particular attention is given to the Marconi station as the publication of this book coincided with the Centenary of the building of this station and its important achievement, in January 1901, of the first over-the horizon wireless telegraphy. The following chapter presents some of the early activities of Marconi.

The cliff scenery

Housel Bay is characterised by beautiful cliff scenery which remains almost totally unspoiled. Bumble Rock, shown in Plate 2, stands guard in the sea at the westerly end of the bay, while to the east are the high cliffs of Pen Olver (Plate 3) and Bass Point. At the centre is Housel Cove (Plate 4) with a very pleasant sandy beach and the cliffs to either side are inhabited by gulls, fulmars and other birds.

A spectacular event occurred in 1847 when without warning the roof of a cave known as Daw's Hugo, near the western end of the bay, collapsed into the sea thereby forming a deep hollow in the ground at a short distance from the cliff. This hollow, seen in Plate 5, is still there and has become known as the Lion's Den; it is now a major feature of the landscape and may be examined by peering carefully over its edge.

The year 1894 saw major new construction at the bay with the building of the Housel Bay Hotel (Plate 6) and its associated access road known as Housel Bay Road. The hotel is still the largest building in the bay, and is in an excellent position with marvellous views by day and night.

Between Pen Olver and Bass Point is Polledan Cove, once known as Trecrobin and then Belidden Cove and the area of ground above it is referred to in some documents as an ancient amphitheatre of Belidden with a suggestion that this was possibly a Druidic temple. There is no proof of this, so the individual is left to decide for himself what to believe, though the change of name on several occasions indicates perhaps that there was something special about this place.

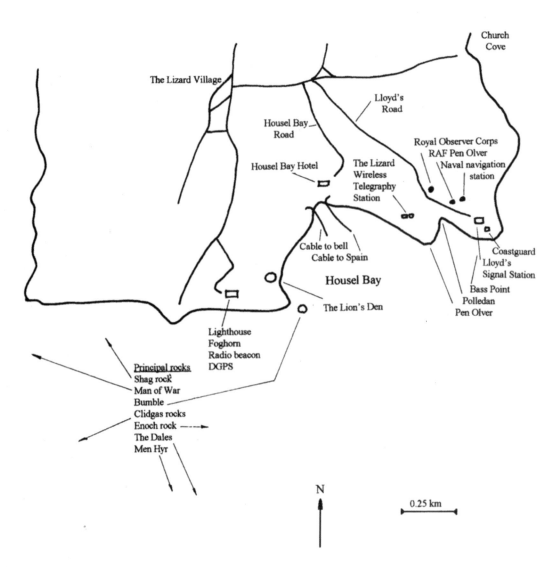

Figure 1. A sketch map showing Housel Bay

Marconi's Early Achievements

Guglielmo Marconi was born at Bologna, Italy on 25th April 1874 to an Italian father and an Irish mother. He had his first introduction to England at the age of five when for two years he attended school at Bedford, but spent the remainder of his childhood and youth in Italy and then studied physics at Livorno. In his late teens he was trying to decide into which branch of physics he should devote his future studies, and it was then that he met Professor Righi who was undertaking scientific experiments on the newly discovered electromagnetic radiation. This radiation had been predicted mathematically in 1864 by a Scotsman, James Clerk Maxwell, and a number of scientists had set up laboratory experiments in the following years to try and produce and detect it. A German, Heinrich Hertz, first achieved this in 1888. Professor Righi was one of a number of experimenters who then continued with scientific studies into electromagnetic radiation.[1]

Experimenters overwhelmingly devoted their activities to scientific aspects of the radiation, but the more Marconi (for such we will abbreviate his name) thought the more the idea formed in his mind that here was something that could be turned into a means for practical communication and he decided to undertake his own experiments. In 1894, using a spare room at Villa Griffone where he then lived near Bologna, he set-up the relatively simple apparatus that was necessary to produce and detect the radiation, and with his transmitting equipment in the same room as the receiver he was soon achieving similar ranges of detection as the scientists in their laboratories.[1]

However, a communication system if it were to achieve a practical application would need a considerable increase in the range of detection from the few metres then possible and this subject was uppermost in Marconi's mind. He was also undertaking quite separate studies on the detection of thunderstorms and for this he used an elevated wire. One day he had the inspiration to join one of the terminals of his electromagnetic radiation receiver to this elevated wire and the other terminal to a plate in the ground and he raised a similar elevated wire and installed a ground plate for the transmitter. This produced a dramatic improvement in performance and for the first time the transmitter and receiver were not constrained to the same room.[1]

Almost from the beginning Marconi included devices in his apparatus that would allow the control of transmissions according to the Morse Code in order to permit telegraphic operation. He experimented with various different heights of his elevated wires and with a variety of conductive sheets, cylinders and cubes connected to them. He included an improved detection device known as a coherer in his equipment which other experimenters had also introduced and also other devices to improve performance and to permit telegraphic operation. By 1895 he had extended the detection range to 2.4 km (1.5 miles), and it was then that he approached the Italian authorities seeking support for his work but was met with a lack of enthusiasm. So at this stage, with no support at home, in February 1896 at the age of 21 he came to Britain, the major trading and maritime nation at that time. It was here that he considered he would be more likely to find support.[1]

Marconi continued his work in Britain, and took the following as priority objectives.
- The achievement of longer ranges of operation
- The demonstration of his equipment to potential customers

- The development of his apparatus into practical wireless telegraphy equipment in order that he might set-up a development and manufacturing company and sell equipment.[1]

His first approach in Britain on 20th May 1896 was to the Secretary of State for War to whom he proposed an application of wireless telegraphy for the remote steerage of torpedoes and boats, and he offered a demonstration but this was not taken up.[2] In June 1896 he was granted the world's first patent on wireless telegraphy, British Patent No. 12039, having made his first application in March of that year. He gave demonstrations of his equipment on Salisbury Plain in 1896 and these aroused the interest of several departments which attended, including the Post Office, Army, Navy, Lloyd's and Trinity House; these all became potential customers. In July 1897 Marconi formed a company to develop equipment commercially, which was registered as "The Wireless Telegraph and Signal Company Limited".[1]

By 1897 the achievable range with Marconi equipment had increased to 7 km (4.4 miles), and then in 1898 to 23 km (14 miles) and in 1899 to 51 km (32 miles) across the English Channel and in naval demonstrations. In these latter demonstrations there was sometimes reception at longer ranges up to 140 km (88 miles) giving an early indication that perhaps the horizon was not a limit for wireless.[1] All of these great extensions in range were achieved in particular by improvements in aerials and coherers and by the application of some elementary circuit tuning.[1,3]

Marconi was not alone in this development work, but he was especially well adapted to it and consistently achieved results which others could not match. In February 1900 the name of the company was changed to "Marconi's Wireless Telegraph Company Limited".[1]

By 1900 Marconi had made good progress on two of the objectives he had set himself in 1896. Firstly the operating range of his wireless telegraphy equipment had been very significantly extended. He had achieved ranges of about 51 km (32 miles) with indications that this was not a limit to possible operation. Secondly this equipment had been demonstrated in trials to a variety of potential customers. Buyers of equipment were, however, few and far between, and the only orders of commercial significance had come from the Navy. The company needed to improve its sales and financial position, and it decided that in order to create customers and to overcome legal obstacles it would lease wireless telegraphy equipment to ship owners and provide operators on the ships, which installed the equipment. To support this activity the company would build a number of coastal stations, which would be able to communicate with these ships. A separate company was formed in 1900 to handle the associated mercantile marine business; this was the "Marconi International Marine Communication Company Limited".[1] The story of one of the coastal stations built in 1900 is told in a later chapter.

References

1. A History of the Marconi Company. W J Baker. Methuen and Company Limited. 1984.
2. The Early History of Radio - from Faraday to Marconi. G R M Garratt. Institution of Electrical Engineers in association with the Science Museum. 1995.
3. Marconi Wireless Telegraphy. G Marconi. The Engineer. December 17th 1909.

The Lizard Lighthouse

The first lighthouse at The Lizard

An authorisation to set up a light above the cliffs at the western end of Housel Bay to warn ships of the dangerous coast had first been given in 1570, but nothing was built. Then in mid 1619 Sir John Killigrew put forward a new request to set up such a light, and despite arguments that it would provide too much of a guide to pirates and enemy forces, and opposition from Trinity House who stated that the light was unnecessary, the authority was given and Sir John Killigrew, at his own expense, built a tower on his own land, and the fire on it to give the light was lit in December 1619. A condition was imposed that the light had to be extinguished if an enemy force was deemed to be approaching. The locals too had opposed the light and they created difficulties during its construction because they foresaw a reduction in their shipwreck spoils.[1]

The only possibility of a financial income to pay for the costs of building and operating the light was by voluntary contributions from ship owners and these were not forthcoming. The light, therefore, became a financial burden on Sir John Killigrew, so much so that he sought the right to charge passing ships, and was granted this right at one half pence per ton. However, this increased the number of people opposed to the light as the ship owners, despite the advantages of the light to their ships, now objected to the levy. With the opposition of the ship owners, Trinity House and the locals, and with difficulties in collecting dues and the fear of pirates, Sir John Killigrew was forced to terminate the enterprise and the light was extinguished after only four years and its tower became derelict. Sir John Killigrew was himself suspected of involvement in piracy and one of his interests in the light was said to be to help his pirate ships.[1]

The second lighthouse

Meanwhile, with the coast now devoid of light, The Lizard rocks at night continued to take their toll of shipping, and by the mid eighteenth century it was considered that something had to be done. In 1748 permission was given to construct a new lighthouse, with Trinity House now supporting the case. Trinity House provided a lease to a Mr Thomas Fonnerau who actually built the new lighthouse at the same site as the former one and it came into operation in 1752; Mr Fonnerau paid a leasing fee to Trinity House and in turn collected fees from ship owners. Since that date the lighthouse has provided a nightly light, warning of the presence of rocks and coast and providing an aid to navigation, yet at the same time giving a "welcome home" to returning travellers. The light was obtained by the burning of coal. It had been considered that four spaced fires should be used in order that ships would be able to recognise the location, but in the event two separate towers were built with a coal fired brazier on each (see footnote). The two towers 66 metres (216 ft) apart enabled The Lizard Lighthouse to be distinguished from the lighthouse on the Isles of Scilly which had one tower and the one on the Casquets Reef off Alderney which had three. The galleries of the towers at The Lizard Lighthouse were 12.2 m (40 ft) high above the ground and 67 m (220 ft) above mean high water level.[1,2]

The problems of keeping the fires burning on a wild night and the lights bright can only be imagined. Between the towers was a small building in which a watchman with a view of both lights was responsible for alerting the fire-keepers by a blast on a cowhorn if the fires grew dim. It would seem that the watchman himself sometimes had to be woken, as for instance when on one occasion a government packet ship fired a cannon when the lights were dim.[1]

Improvements by Trinity House

Trinity House took over responsibility for the lighthouse in 1781 when Mr Fonnerau's lease had been ended, and in 1811 it was decided that a much improved light could be achieved by a change to oil lamps and these were installed by 1812 using the same two towers as before. During the modifications temporary lanterns were raised on poles. Major other work was undertaken at the same time including houses, offices and a long passage which enabled keepers to reach each tower without having to venture outdoors. The new Argand oil lamps provided a wide stationary beam from each tower with the beam from one tower fully overlapping that from the other such that both lights were visible over a 230 degree wide sector of the sea. The light from each tower was in fact provided from an array of 19 stationary burners each with its own 53 cm (21 inch) diameter polished parabolic reflector; the nineteen burners were located inside a plate glass lantern. These lights were a great improvement and were clearly visible in fine weather from the horizon at 34 km (21 miles) and the usefulness of the two lights as leading lights to guide ships to avoid distant headlands and rocks to the east and west was noted.[1]

The year 1878 saw another change when the oil lamps were replaced in their turn by brilliant electric arcs, again retaining lights on the two towers with stationary beams.[1,3] These required the generation of electricity at the lighthouse and this was provided by a generator seemingly driven by two hot-air engines, and an engine-house was built to accommodate this machinery; one of the hot-air engines was replaced by an oil engine in 1895 and the other in 1907. The arc lights were very bright, so much so that they lit up the sea and country within the confines of then-wide beams, and vessels at sea were clearly visible. In 1883 a second electric generator was installed, this being one that had been exhibited at the Paris exhibition of 1881; a generator of this vintage has been retained at the lighthouse as a museum piece.[1] The 1891 census record shows that there were 39 people, including children, living at the lighthouse; seven of these were light keepers of whom one may have been on leave from an off-shore lighthouse and there was one engineer.[4] A photograph of the lighthouse when it still had two lanterns is shown in Plate 7.

A major change was made in 1903 when the two-light wide-beam operation was terminated and since then there has been a single light on one of the towers.[1,3] The light beam was then narrowed to one degree wide by a glass lens in front of the light source, this lens being mounted on one face of a square frame, with the other three faces each holding a similar lens. There were, therefore, four separate lenses angled in azimuth at 90 degree intervals, and the whole assembly of these and the frame weighing 4100 kg (4 tons) floated in a bath of 363 kg (800lb) of mercury (see footnote 2) so that the torque to turn it was quite small and was provided by a slowly descending 356 kg (7 cwt) weight suspended in a channel passing down through the centre of the tower. It was a task of the keepers to rewind the weight every half-hour. There were therefore four narrow beams at 90 degrees to each other, which scanned the sea in turn as the assembly rotated at one revolution every 12 seconds. Thus a distant observer would see a short flash of light every 3 seconds, and this is the characteristic of The Lizard light. An electric arc light was still used and the intensity of the focused beam was equivalent to 12 million candle power making the lighthouse the most powerful in the world and visible from the horizon whatever the height of the ship; this was typically 50 km (31 miles) for a 15 metre high ship. Each of the four lenses were made of 91 prism shaped glass pieces, and the technical description of each lens is dioptric at its central regions where the light from the arc is refracted into the beam, and catadioptric for the outer parts, where the high angles from the arc prevent simple refraction into the beam and the light is both refracted and reflected internally inside each of the outer prisms.[1,2]

A change from the electric arc to a 3 kW electric filament lamp was made in 1926 and the beam was thereby reduced to 5.25 million candle power. A spare lamp was located near the one in operation and if one failed the other was automatically moved into position and brought into operation. If both lamps or the electric supply failed then an acetylene gas burner was automatically lit and focused. Mains electricity was brought-in in 1950 with the generators retained as a standby assurance, and turning of the lens assembly was made electric at this time and the descending weight system terminated.[1]

Whilst the lighthouse is well known as a white building, a painting by John Edgar Piatt shows it in camouflage colours in 1942.[5]

Automation

In 1998 the lighthouse became automatic and the electric lamp was reduced to 400 W, the lens assembly of 1903 still being retained. Now no one lives there although the houses are used as holiday lets. Being a shore station and open to the public at a popular holiday location, the station is visited by many people each year and guided tours of the tower and engine house are provided. Prior to automation the lens assembly was stationary during the day and curtains were drawn to keep-out the sun's rays which, focused by the lens, would have damaged the equipment. Now with automation the curtains are no longer used and the lens assembly is kept rotating during the day to avoid damage by the sun. Should the electric lamp fail then a standby lamp is brought automatically into place and operation. Should the whole light or lens fail then there is a separate and small standby battery-operated low-power light mounted outside the lantern and this then operates as an emergency measure until the main light is restored.[2,6]

Both octagonal towers and the houses and engine house still stand on this the most southerly lighthouse on mainland Britain and at the western gateway to the English Channel, and the lighthouse is now a Grade 2 Listed Building still owned by Trinity House and meticulously maintained. Two recent photographs are shown in Plates 8 and 9.

Footnote 1

Books and articles contain a variety of stories in relation to the four towers, some say that four were built and used,[7,12] one that four were built and used but reduced later to two,[8] others that four were considered but only two built.[1,9,10,11] The more detailed research of Reference 1, however, supports the position that whilst four towers were considered only two were built, and this is confirmed in Reference 2 by Trinity House which stated at the time that maintenance and coal costs would be excessive with four towers and that it would only approve the request for the right to establish a light if the construction was with two lights.

Footnote 2

The weight given for the lens assembly and mercury are included as they are given in several references which all seem to rely on one source. The weights, however, according to the laws of physics, are incompatible and more likely figures are considered to be in the order of 1400kg and 1500kg respectively.

References

1. Cornish Lights and Shipwrecks. Cyril Noall. 1968.
2. Papers and letters from Trinity House.
3. The Diaries of Mr T S Hendy. 1874-1930. By kind permission of Mrs E Hendy of The Lizard.
4. The National Census Record. 1891.
5. Regional Newsletter of the National Trust. Spring/summer 1997.
6. Guided tours of the lighthouse by the keeper, Mr Eddy Matthews.
7. Lighthouses, Their History and Romance. W T Hardy. 1893.
8. Guide to The Lizard and Kynance. J H Miners. 1933.
9. A View from the Sea. R Woodman. Century Publishing.
10. The World's Lighthouses before 1820. D A Stevenson. Oxford Publishing. 1959.
11. Lighthouses. Their Architecture, History and Archaeology. D Hague.
12. A History of Lighthouses. Patrick Beaver. Peter Davies Publishing.

Additionally a number of other books over the years have included sections on The Lizard lighthouse and some of these are listed below.

a. A Week at The Lizard. Rev C A Johns. Published in 1848, 1863, 1874 and 1882 by The Society for Promoting Christian Knowledge.
b. Lighthouses. W H Davenport Adams. 1898.
c. British Lighthouses. J P Bowen. Longmans Green and Co. 1947.
d. Cornish Sea Lights. D Mudd, Bossiney Books. 1978.
e. Cornwall's Lighthouse Heritage. M Tarrant. Twelveheads Press. 1990.
f. Trinity House. The Super Silent Service. M Tarrant. Gomer Press. 1998.

The Lizard Signal Station

The electric telegraph

Overland messages in the early nineteenth century were carried primarily by hand, often on horseback or by coach, though there were a few fine-weather special communication systems. For instance, a string of bonfires to warn of the threat of invasion or a chain of semaphore stations as was used by the Admiralty between London and Portsmouth. Ships were the only means for carrying messages overseas.

Electricity was to play a large part in changing all this. The ability to charge such materials as amber with static electricity had been known for thousands of years, as had some of the effects of magnetism. Then at the end of the eighteenth century the discovery of particular properties of certain materials permitted scientists to make electric cells and batteries of cells which would provide a steady flow of electric current in a metal and this made it possible to undertake a range of electrical experiments in laboratories. Several scientists in the early nineteenth century then established many of the basic rules of electric circuits and current flow. Alessandro Volta, Hans Oersted, Georg Ohm and Michael Faraday were well to the fore in this activity and have left their names in the rules and terms used in electrical engineering. Continuing experiments demonstrated the ability of an electric current to produce a magnetic field, and also of a moving magnetic field to generate a current; these and the effects of one electric circuit on another were to become important features of electrical engineering.

This scientific work led to practical applications for electricity, the first of these being the electric telegraph, which by sending electrical impulses along a metal wire according to an agreed code permitted messages to be conveyed from one place to another at great speed. The expanding railway network in the 1830s provided the impetus for the application of this new technology. Metal wires hung on poles alongside the track provided the means to convey electrical telegraph information between railway stations at a speed faster than the train and in all weathers by day and night, thereby assisting a safe and fast train service. By 1850 there were over 3500 km (2200 miles) of telegraph wires associated with British railways. The Admiralty showed an interest by 1844 in sharing the railway telegraph from London to Portsmouth, and Lloyd's were soon using the electric telegraph to report ship arrivals. Mines, newspapers and the police also adopted the system.[1]

The electric telegraph was introduced to the general public, and the usage grew steadily as a network of wires and telegraph offices was constructed across the land. In Britain this was

13

organised by public companies until 1870 when the Post Office took-over responsibility; the charge for 20 words was a shilling (5p) up to 160 km (100 miles) and two shillings (10p) over 480 km (300 miles); the address being free. For this a message could be handed-in at a local telegraph office and be received at another office many miles away in minutes and then delivered by a telegraph boy on a bicycle.[1]

A signal station by G C Fox and Company

Ships trading across the world were mainly wind driven in the early nineteenth century, and as they approached The Lizard many would have had no communication with their homeports or their owners, perhaps for months. Their only method of navigation when out of sight of land had been with the compass, sun and stars, and the owners would be keenly awaiting their return and anxious for their safety. They would be wondering what cargoes they carried and would want to make arrangements for berthing and unloading and for the dispersal of the goods, and then for reloading. At Falmouth a keen watch was kept for the appearance of any ship in the bay, and then gigs would race to be the first to reach the ship and put a pilot aboard.

One company of ship owners and agents at Falmouth was G C Fox and Company, and it was to make good use of the electric telegraph to obtain advance information on ships before they had reached Falmouth Bay.[2] The overland electric telegraph had reached as far as Falmouth by 1857, and put Falmouth in almost instant touch with the towns of the rest of the country.[1] G C Fox and Company considered that if the electric telegraph were extended to The Lizard then it would be able to report the presence and progress of passing ships. Further, by signalling to the ships from The Lizard, information could be conveyed to and from the owners.[2]

G C Fox and Company, in the early 1870s, considered building a signalling station in a field near the lifeboat house at The Lizard Head, the lifeboat house then being at the top of the cliff where there is now a small car park, cafes and shops. In the event in 1872 they built their signalling station on the cliff top at the eastern end of Housel Bay at Bass Point (once known as Beast Point)3 and constructed a road for access to and from the village. The building known as The Lizard Signal Station had a flat roof on which was erected a mast for signal flags; there were five small look-out windows near the top of the building which gave a good view to seaward through which a watch was kept for ships. With this station the company was able, in fine weather, to record and report the passage of ships which were in sight of land, and if owners wished to communicate with their ships messages could be sent to and from the ships using a recognised array of flags. The station became operational on 1st April 1872, Figure 2 being a copy of an announcement of this.[2,4]

The company had been prepared to lay its own telegraph cable to the station from its offices in Falmouth, but the Post Office declared that it would do this. This part of the construction, however, ran late and for a short period the messages were sent by post or in cases of urgency carried by horse to the nearest telegraph station which was at Helston. Thus for a period the transfer of information between the offices at Falmouth and the station was slow, and the company threatened to claim recompense from the Post Office for this inconvenience. However, the telegraph was on its way and installation to the Signal Station was completed and telegraphic communication with Falmouth and the rest of the country was first achieved

14

on 2nd June 1872, see the announcement in Figure 3. Rooms at the station were leased to the Post Office by G C Fox and Company, and the Post Office set up a telegraph office there and was responsible for operating it and sending and receiving overland messages. The company was now able to convey messages in short time, not only between London and its offices in Falmouth, but also between Falmouth and ships in the bay at The Lizard, providing a great increase in its communication efficiency.[2] Operation was extended to night-time using steam whistles, guns, rockets and by ships displaying geometric arrays of coloured lights, but this was very limited in fog. Figure 4 is a copy of an announcement on this type of operation.[2]

The services of the station were offered to other shipping companies, and this method of communication became a popular 24-hour operation, the telegraph office staying open at all times. Charges were made to other companies for the services provided. In the first nine months 1827 ships were either reported as passing or were signalled.[2] The station, therefore, provided a major step forward in maritime communication.

A rival shipping office at Falmouth, William Broad and Sons, set up another signal station with a building, platform and flag mast just to the east of the G C Fox station, and it entered into competition. Working relations between the two companies were not the best at this time and William Broad and Sons for instance wanted to break through a hedge so that it could have access to the road; however, G C Fox obtained the help of the fishermen from nearby Church Cove and lined them along the hedge in question to prevent any such mutilation. It was not always easy for either station to know which ships wanted to use which station, and there were cases of both stations charging shipping companies. This dual operation caused a certain amount of frustration both with the shipping companies and the stations, with the shipping companies selecting the lower prices when it suited them. In 1874 the two companies entered into discussions on possible joint operation and on 1st January 1875 the companies announced that they would unite their activities at the G C Fox station. The station of William Broad and Sons was then dismantled,[2] and there is now no trace of it. By 1877 more than 1000 ships per month were using the station.[2]

Lloyd's takes over

In December 1882 Lloyd's distributed a document to shipping companies stating that they had either planned or established 27 signalling stations around the coast of Britain and Ireland. Each of these was named and one was The Lizard. Lloyd's would have built their own station at The Lizard in competition with the Fox station and there would then again have been two stations operating side by side; Lloyd's with their wider operation over the country would have been able to under-charge at The Lizard and therefore put the existing Lizard Signal Station into a difficult situation. G C Fox & Company, therefore, in discussions with Lloyd's during 1882, agreed that Lloyd's would take over the signalling and reporting service at The Lizard and lease part of G C Fox's building as their site. The shipping companies were therefore informed that Lloyd's would take on operation of the station from 1st January 1883, which then became known as Lloyd's Signal Station. The telegraph was still operated by the Post Office.[2] In 1904 the station took out a lease on a small portion of land where a mast and stays might be positioned for wireless telegraphy operation, but this mast was never erected.[2]

Some time before 1939, in order to provide better signalling by lamp without interference from the light of the lighthouse, a separate low building was constructed a short distance in front of the station, and this was known as the night box. Another night box was also built in front of this one at a later date.[5] The station continued to be operated by Lloyd's until 1951, from which time signalling and reporting was conducted from the same building by the Coastguard Service on behalf of Lloyd's.[6]

The building always identified itself with large lettering on the walls, and on its western and southern sides in the nineteenth century were the words V R Telegraph Office; later-on the words on the west face were changed to Lloyd's Signal Station and it is these latter words by which the station is now remembered. In the early years of the twentieth century the south face bore the words "E R Telegraph Office". There was no mains water, mains sewage or mains power at first;[5] the evidence of the water spring source, which was used, can be seen partway down the precarious cliff in Polledan Cove.

Whilst this chapter has described the usefulness of this station, problems sometimes arose at sea. For instance, in order to carry out signalling, ships approached closer to the coast than they previously would have done and sometimes had to wait for a message and, in 1913, the 2144 tons four-masted barque *Queen Margaret*, 130 days out of Australia, signalled the station and while waiting for orders drifted onto the rocks and became a wreck.[7,8]

The present building

The station building is still on Bass Point at the end of Lloyd's Road, and is a Grade 2 Listed Building. It is now owned by the National Trust who put it into good repair in 1994 and restored its exterior appearance to that of former times, even to the inclusion of the Lloyd's name. It is now leased as a private dwelling and the night boxes have also been converted into living accommodation. Photographs of the building and the night boxes as they are now are given in Plates 10 and 11.

References

1. Electric Telegraph, a Social and Economic History. J L Kieve. David and Charles. 1973.
2. The historical archives of G C Fox and Company. By kind permission of Mr C L Fox.
3. Ordnance Survey map. 1813.
4. A Week at The Lizard. Rev C A Johns. Third and fourth editions.
5. Fox's signal Station, Bass Point, The Lizard. S B Smith. 1992. Paper to the National Trust.
6. Cornish Lights and Shipwrecks. Cyril Noall. 1968.
7. The shipwreck records of Mr A E (Bert) Jane. By kind permission of Mr A A Jane.
8. The diaries of Mr T S Hendy. 1874-1930. By kind permission of Mrs E Hendy.

Falmouth 2nd April, 1872

Sir,

We beg to inform you that our
LIZARD SIGNAL STATION is now open, and we are
prepared to signal all Vessels passing that Point and
communicate with Owners, Merchants & others interested

The Station is erected on Beast Point 3/4 of
a mile East of the Lighthouses and is on the highest
Point in the District. The Post Office Department
have undertaken to erect and maintain a Telegraphic
Wire from this Signal Station direct to Falmouth
but pending the completion of this Wire, we purpose des-
patching all important messages by man and horse to
the nearest Postal Telegraph Office, at Helston (about
1 hour distant) and from that place by the Public Wires

We are, respectfully,

GC Fox & Co

Figure 2. An announcement by GC Fox and Company on the opening of their signal station at
The Lizard

From the archive of GC Fox and Company

COMPLETION OF LIZARD TELEGRAPH.

Falmouth, 4th June, 1872.

Sir,

We beg to advise you that Government has opened a Postal Telegraph Office in our Signal Station at the Lizard, so that we are now prepared to report by direct telegraph all vessels signalled off that point.

We shall be glad to receive your instructions about any vessels in which you may be interested, and would suggest that you inform your Captains of the opening of this Signal and Telegraph Station.

We are

Yours respectfully

G. C. FOX & Co.

Figure 3. An announcement by GC Fox and Company that the electric telegraph had reached The Lizard Signal Station.

From the archive of GC Fox and Company

LIZARD SIGNAL STATION

NIGHT SIGNALLING.

The Undersigned have now made arrangements to keep a look out **BY NIGHT** as well as by day, for such Mail and other Steamers as they may be requested to watch for, and to telegraph immediately the passing of any that signal, to owners and others interested.

The Station is **a White** Building, **30** feet high, situated on Beast Point, three-quarters of a mile east of the Lizard Lighthouses, **and** has a flag-staff attached to the flat roof for day signalling.

G. C. FOX & Co.

Falmouth, November, 1872.

Figure 4. An announcement of night-time operation at The Lizard Signal Station.

From the archive of GC Fox and Company

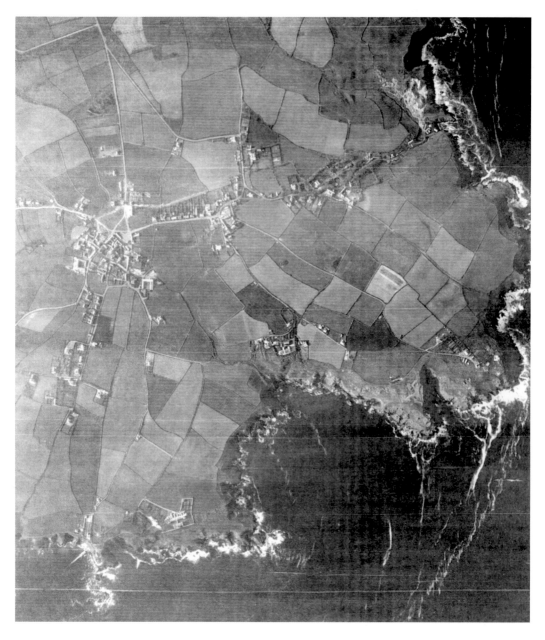

Plate 1.   An aerial view showing Housel Bay and The Lizard Village in 1946

Plate 2.   Bumble Rock

Plate 3.   Pen Olver

Plate 4.   Housel Cove

Plate 5.   The Lion's Den

Plate 6. Housel Bay Hotel

Plate 7. The Lighthouse when it had two lanterns

The Direct Spanish Telegraph Cable

Whilst the electric telegraph had been spreading extensively over land, the possibility of laying cables under water to provide telegraphic communications with countries overseas was being investigated. By 1850 such a submarine telegraph cable had been laid between England and France and, although this was only short-lived, improved technology enabled more reliable cables to be laid, and by 1866 even trans-Atlantic telegraph operation was achieved.[1]

In 1872 there were plans and negotiations to lay an undersea telegraph cable between Spain and England, though by October of that year, even while the cable was being loaded into the laying ship, the landing site and terminal for the cable in England had not been decided. However, the arrival of the overland telegraph cable at The Lizard Signal Station earlier in the year attracted the attention of the cable company, and a decision was made to land the cable at Housel Cove and terminate it in The Lizard Signal Station. On 30th November the cable-ship Dacia set sail from Algorta near Bilbao in Spain, where one end of the submarine cable had been brought ashore and, paying-out cable, reached Housel Bay on 4th December where the cable was cut and buoyed some 16 km (10 miles) from the shore.[2]

The ship then moved closer to the beach, and, with the help of about 100 men and The Lizard and Cadgwith seine boats, some 365 m (1200 ft) of a larger armoured cable was pulled from the ship and supported on floats. The end was brought to the beach where a party of 50 men hauled the shore-end of the cable up the cliff and the end was made fast. This was quite a feat, as the 67 m (220 ft) of the very stiff cable up the cliff alone weighed about 720 kg (1600 lb). The ship then proceeded back to the buoy paying out this armoured cable, which was then spliced to the lighter cable that had been buoyed earlier and the join was then lowered to the sea bed. It had been a busy day in the bay, and an artist as shown in Plate 12 has recorded the event.[2] A shallow trench was cut in the side of the cliff into which the cable was laid and secured by iron cleats and concrete, while rocks and boulders on the beach were moved to provide a clear way for the cable to be buried in the sand.[2]

The cable end at the top of the cliff was connected to a lighter cable, which was laid in a 30 cm (1 ft) deep trench along the cliff top and terminated in a room which had been rented in The Lizard Signal Station. The other end of the cable in Spain was joined to a section of overland cable and this was led in a trench to its termination point at Bilbao. The overall length of cable from Bilbao to The Lizard Signal Station was 877 km (550 miles).[2]

When all was complete the cable was tested electrically on 13th and 14th December by Sir

William Thomson, who in 1892 became Lord Kelvin, and declared to be in "perfect electrical condition". The cable entered service on 31st December, and was placed under the ownership of a new company, The Direct Spanish Telegraph Company.[2] The Post Office overland telegraph cable from The Lizard Signal Station enabled messages to be transferred to their destinations with minimum delay. The cost to the user of the submarine cable was 8 shillings (40p) for 20 words.[2]

This submarine cable operated satisfactorily for several years, though the effects of winds and tides were considerable and led to some interruptions in service. In 1884 a new cable was laid some 6 km (4 miles) to the east into Kennack Bay, near the village of Kuggar, where the long sandy beach was more favourable for the well being of the cable. This cable terminated in a specially built hut by the beach, and a landline connected direct to Falmouth. The cable at Housel Bay was then taken out of service, but a landline was laid from Kennack to The Lizard to provide emergency cover to the inland telegraph service as the Kennack/Falmouth landline was damaged on several occasions.[2,3,6]

The armoured end of the undersea cable at Housel Bay can still be seen at the eastern side of the beach on its meandering track up the cliff face and on the beach in winter when the tides have removed much of the sand. Whereas an overland telegraph cable suspended from insulators on poles was typically quite small in cross section, a cable to be laid on the sea bed had to have an insulation layer to isolate it from the sea; it also needed armouring outside the insulation, especially near the shore, to protect it from sea animals and from the motion of sea, rocks and sand and from fishing tackle. This necessitated a cable at the shore end of 51 mm (2 inch) diameter, see Figure 5. The iron cleats and concrete securing the cable are still evident on the cliff, and the construction of the cable may be examined where it has been fractured part-way up the cliff revealing the inner copper conductor, gutta percha insulation and strengthening layers of iron wires. The parts of the cable on the beach which can only be seen at times in the winter have some well preserved inner portions of the cable; the gutta percha being in good condition. Some copies of photographs showing the cable as it is now are given in Plates 13 to 16.

References

1. Girdle round the Earth. H Barty-King. Heinemann. 1979.
2. Submarine Cables between The Lizard Peninsula and Spain. D M Todd. Unpublished report.
3. Reminiscences of a Postmaster of Falmouth 1856 to 1896. Journal of the Cornwall Family History Society. December 1994.
4. Kelly's Directory of Cornwall. 1873.
5. A Week at The Lizard. Rev C A Johns. Editions 3 and 4.
6. Correspondence in volume MS 31674/3 at the Guildhall Library, London

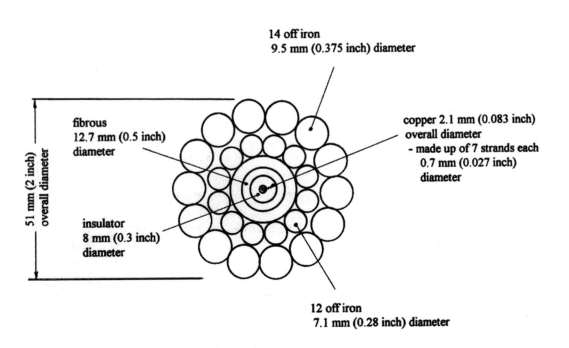

14 off iron
9.5 mm (0.375 inch) diameter

fibrous
12.7 mm (0.5 inch)
diameter

copper 2.1 mm (0.083 inch)
overall diameter
- made up of 7 strands each
0.7 mm (0.027 inch)
diameter

insulator
8 mm (0.3 inch)
diameter

51 mm (2 inch)
overall diameter

12 off iron
7.1 mm (0.28 inch) diameter

Figure 5.   A cross-section of submarine cable

The Foghorn

The development of the electric telegraph, both overland and undersea, had provided enormous improvements in communication by the second half of the nineteenth century but had still left ships relying on visual methods. This was convenient for much of the time but was completely inadequate during fog, and in such conditions many ships never saw land until it was too late to avoid hitting rocks or cliffs. Concern had been expressed for many years at the large number of shipwrecks at The Lizard during fog, and Reference 1 states that a device called a steam gun to provide an audible report every five minutes had been suggested before 1848 but nothing was done.[1]

By 1873 there were plans for muzzle-loading cannon to be used as a fog signal; these would have been located in a purpose-built house with a separate magazine but these were not constructed.[8]

Then in 1878, when major changes were being made at the lighthouse, Trinity House installed a new device - the foghorn - which at regular intervals during fog blasted an audible signal out to sea in the hope that ships would be able to determine from which direction it was emanating and thus steer clear of it and the coast.[2,3] This foghorn was located at the front of the new engine house at the lighthouse, with a single transmitting horn laid horizontally on the low roof. This horn could be turned to face into the wind.[8] The foghorn was powered by compressed-air provided by a compressor driven by a hot-air engine, the compressed air being held under pressure in large storage receivers. (Thus both light and sound were produced by heat engines here in 1878). At first the foghorn was sounded every 5 minutes when foggy, but this was changed to a high/low sequence every 2 minutes, and by 1907 had changed again to a long/short sequence every minute. At this time the single horn was changed to two horns which protruded vertically through the low roof and then bent to the horizontal towards the sea as shown in Plate 17. The long/short sequence was with a 7 second blast, then a 2 second silence followed by a 2 second blast, and then a 49 second silence, this sequence being retained up to 1998. The audible range could be up to 16 km (10 miles) in good conditions. An oil engine replaced the hot-air engine which drove the compressor in 1901.[2,4] It has been recorded that the foghorn sounded for about 540 hours annually.[5]

The benefit of the foghorn would seem to have been very limited as examination of records shows that fog continued to be the most frequent cause of shipwrecks and such shipwrecks continued even including a cargo-liner, the *Suevic*, homeward bound from Australia which came to grief on the rocks in fog just to the west of Housel Bay in 1907. The local lifeboats rescued 456 people on that day. This still remains as a daily record for the Royal National Lifeboat Institution.[6]

The different notes of the foghorns of passing ships were clearly audible on shore until a few years ago, and even though these are now seldom heard The Lizard foghorn has always continued. It is perhaps surprising that more effective methods to aid survival in fog were not developed in the 19th century.

To enable the station to be automated in 1998 the compressed-air foghorn was taken out of service though it was still kept in working order for some time and sounded on occasions but it is now silent. A new electronic foghorn has now been installed, and this is located on the gallery of the same tower as the light; see Plate 18. This foghorn operates at five times higher audio frequency than the previous one and sounds in fog every 30 seconds; it is set into operation automatically by a water particle sensor on the tower.[4,7,8]

It is gratifying to see that the automation of the lighthouse has enabled the optical and audio equipment in use before 1998 to be retained or used so that succeeding generations of people can see directly the engineering achievements of a hundred and more years ago.

References

1. A Week at The Lizard. 1848. Rev C A Johns.
2. Cornish Lights and Shipwrecks. Cyril Noall. 1968.
3. The diaries of Mr T S Hendy. 1874-1930. By kind permission of Mrs E Hendy.
4. Papers from Trinity house.
5. The Yearbook of The Lizard Women's Institute. 1965.
6. The shipwreck records of Mr A E (Bert) Jane. By kind permission of Mr A A Jane.
7. Guided tours of the lighthouse by the keeper, Eddy Matthews.
8. Lost Sounds - the story of coast fog signals. Alen Renton. Whittles Publishing, 2001.

The Lizard Wireless Telegraphy Station

"Guglielmo Marconi whose pioneer work in wireless telegraphy for the safety of all seafarers was furthered in this building during the first years of the (twentieth) century"

The birth of wireless

The importance of the early scientific work on electricity has been mentioned in a previous chapter. This led to practical applications of electricity, and in Housel Bay we have already seen before the twentieth century the overland electric telegraph, the underwater electric telegraph and the electric light.

As mentioned earlier, during the 1860s a mathematical physicist, James Clerk Maxwell, whilst endeavouring to understand some of Michael Faraday's experimental findings with electric and magnetic fields, produced a theory which predicted that it would be possible to generate an electromagnetic radiation that would propagate. He set this down in a paper entitled "A Dynamical Theory of the Electro-magnetic Field" which he read to the Royal Society in 1864.[1]

This in its turn attracted the notice of experimental scientists who designed and set-up experiments in their laboratories in endeavours to produce and detect such radiation. It was not until 1888 that an experimenter by the name of Heinrich Hertz announced that he had produced, radiated and detected electromagnetic radiation. He used electrical discharges across a spark-gap connected to two metal sheets to produce the radiation, and he detected the radiation a few metres away using a wire loop which had a small gap across which he observed sparks when the radiator was operating.[1,2]

Another scientist involved with studies and experiments on electromagnetic radiation was Oliver Lodge, and he may be considered as narrowly missing being the first to both produce and detect the radiation for he was well advanced with his experiments but was beaten to the post by Heinrich Hertz.[1,2]

The experiments of Heinrich Hertz had used a simple wire loop detector, but in 1890 Edouard Branly reminded the scientific world of the property of certain metallic powders to change their electrical conductivity when close to electric sparks, and he undertook considerable research into this.[1,3] Oliver Lodge demonstrated in 1894 with an improved Branly tube that the change in conductivity was caused by the electromagnetic radiation from the spark and gave to the tube the name of coherer after the ability of the particles to cohere when the radiation was present.[1,3] The coherer was a very convenient detecting device, and for the next ten years was to be the primary detector of electromagnetic radiation.

In 1894 Oliver Lodge in co-operation with Alexander Muirhead gave a demonstration of wireless operation between two rooms about 55m (180 ft) apart and also showed how it was possible to code the transmissions; this was the first demonstration of wireless telegraphy, but these two did not pursue this work to an immediate practical application.[1]

A new Marconi station at Housel Bay

The activities of a young Italian who devoted his work to achieving a communication system using electromagnetic radiation have been described in an earlier chapter. In 1900 the Marconi International Marine Communication Company Limited constructed a new wireless station on the cliff top on the eastern side of Housel Bay, called The Lizard Wireless Telegraphy Station. This was one of the first group of eight coastal stations to be built in 1900/1901 which operated on ship to shore wireless telegraphy and which in these early years earned much needed money for the company. So in 1900 for the first time ships at sea which were equipped with the wireless apparatus were able to communicate with each other and with the shore, in a telegraphic form in all weathers by day and night. The Lizard station, however, had other roles, which it was to undertake to further the company aims in development and application of wireless telegraphy; these will be explained in later paragraphs.[3]

The station was made up of two wooden huts and a 48.8 m (160 ft) high mast which held aloft long wire aerials. One hut housed the transmitter and receiving equipment, and the other provided living accommodation for the operators. The Lizard station was operational by January 1901, and the huts and aerial mast may be seen in two Marconi photographs estimated to date from 1903, and these are reproduced in Plates 19 and 20.[3,4,5]

There were three reasons for choosing Housel Bay for the site of a Marconi station.

a.   It was a convenient site on which to construct a working coastal station as many ships from around the world were passing every day.
b.   The company was also constructing a high-power and large-aerial station at Poldhu some 10 km (6 miles) away, and The Lizard station would be able to test the Poldhu transmissions during the development of the Poldhu station.
c.   The company was developing circuit components, known as jiggers, in order that several stations might be able to operate in close proximity, and The Lizard station would be able to test the effectiveness of these under operational conditions and in close proximity to a high power station.[3,5]

Thus The Lizard station was to combine commercial service with an experimental function at a critical time for the new company.

A description of early wireless equipment

A wireless telegraphy station contained three primary pieces of equipment, the transmitter, the receiver and the aerial.

The transmitter, shown in Figure 6 in its form in the years up to 1900, made use of the characteristics of an electric spark for producing the electromagnetic signal. When the transmitting Morse key was depressed, a battery of electric cells powered the primary winding of an induction coil containing an automatic make and break circuit. The secondary winding of the coil, in which a series of very high voltages were induced by the action in the primary, was connected to two metal balls (the spark-gap), and the high voltages discharged across the gap in a series of sparks. This discharge produced a chain of electromagnetic bursts of power for as long as the key was depressed. Thus a short burst of sparks and radiation would represent a dot of the Morse code and a longer burst a dash, with the periods of no radiation representing the spaces between the dots and dashes. The balls of the spark-gap were in turn connected to the aerial and ground which would radiate the electromagnetic power produced.[3]

The receiver made use of a coherer (see earlier section of this chapter) connected to an aerial and ground, and was only operational when the transmitter was quiescent. The coherer up to 1900 was connected directly to the receiving aerial and to an electric cell and relay, as shown in Figure 7. The circuit was adjusted such that when the conductivity of the coherer changed on receipt of a wireless signal the change in current from the cell caused the relay to close a contact in a secondary circuit which caused another battery of cells to operate a printer. One feature of a coherer was that once cohered it remained cohered, and so a vibrating tapper was made to strike the coherer continually causing the particles to separate when no signal was being received. Thus the printer only operated when a signal was present and, therefore, recorded the series of dots and dashes of the received message which could then be decoded by the telegraphists.[3]

Marconi had demonstrated at an early stage the importance of aerial height to achieve long-range operation, and wire aerials up to 48.8 m (160 ft) in height supported by a wooden mast were in use in 1900. Such a mast as at Housel Bay was constructed in three steps as for a ship's mast. The aerials themselves were simply long wires, one for the transmitter and one for the receiver, with one end of each held aloft by the tall mast, and provided the means for effectively radiating and receiving the electromagnetic signals.

The multiple-station solution

The spark-gap transmitter in its simplest form, whilst it produced the required electromagnetic energy, did so in the early circuit configurations (Figure 6) over a wide frequency band. The simple receiver in its turn (Figure 7) was open on a wide frequency band, and so if several nearby transmitters were in operation all signals would be heard by the receiver and the messages would be intermixed and unintelligible.[3]

To overcome this problem, Marconi developed what was then called a jigger, which was a transformer and capacitor combination which for the case of the transmitter was inserted between the spark-gap and the aerial, and was tuned to act as a filter to the required transmission frequency by selecting the associated values of inductance, coupling and capacitance, and the energy of the transmitter was thereby concentrated into a much narrower band than before. The arrangement is shown in Figure 8. A similar jigger was included between the receiving aerial and the coherer, see Figure 9, and this could be tuned to select whatever transmitting station was to be received and others were rejected.[3] The use of jiggers overcame a further problem in that the spark-gap and the coherer were not well matched to the aerials when connected directly, but the jiggers enabled matching to be achieved and the performance of both transmitter and receiver were improved by this move. These jiggers and their application were patented by Marconi in 1900 in patent number 7777, known as the four sevens patent.[3] The effectiveness of the jiggers in 1901 at The Lizard station was demonstrated in their ability in cutting out interference from a French station.[6]

Technical and other details of The Lizard station

The Lizard station, located on a 2500 square metre site, was contained in the taller but smaller two-roomed wooden hut[7] measuring 6.5 x 3.5 m (21 x 11 ft). This is confirmed by the results of a survey by the Cornwall Archaeological Unit in 1997 which display shadow markings on the wall[8] coinciding with items in a photograph reproduced in Plate 21 which shows the interior arrangement of equipment at the station. The remains of an aerial feeder which still exists in the wall also confirm the use of this hut as the operating building.

In the early days of the station a battery of dry cells powered the spark-gap transmitter. The receiver used a coherer. Jiggers were available for both transmitter and receiver. The photograph in Plate 21 of the inside of the hut shows that the transmitter and the receiver were both located in one room. The transmitter is at the right end of the bench with the induction coil and spark-gap at the extreme right and the operating key in front. Just to the left is a condenser for the transmitter made-up of a bank of six Leyden jars and the associated jigger is in the box above. On the left of the bench are two coherer receivers and in the centre is the printer with battery alongside.[4]

The second hut measuring 9.7 x 3.5 m (32 x 11 ft) is also in the Marconi photographs of about 1903 and was used as living accommodation.

The height of the aerial mast at 48.8 m (160 ft) has been determined by scaling the mast in old photographs against the cliff and buildings and from data from the Marconi archive at the Bodleian Library. At the top of the mast is a 5 m (16 ft) gaff with an aerial wire suspended from each end.

The call sign of the station was LD,[7] presumably from letters of The Lizard.

The range argument and "Marconi's First Great Miracle"

The scientific community in 1900, familiar with studies in optics, considered that the operating range of wireless would be restricted to the horizon, as was optical radiation. They were, after all, both electromagnetic radiation but at different frequencies. Beyond the optical horizon the curving surface of the Earth would put one station in the shadow of the Earth as viewed from the other station, and so it was said that no reception would then be possible. The maximum range at which two wireless stations can just be in direct line of sight with each other on the spherical Earth is given by the following formula.

$$d^2 = 67.2\,h$$

where the station separation d is in km, and the aerial heights h are in m.[9]

In January 1901, Marconi decided to attempt a long distance communication between two of his new stations. He chose the Isle of Wight and The Lizard, both stations having his latest equipment including jiggers and high aerials. If the scientists were correct then success would only be possible with masts about 1300 m (4300 ft) high at each of these sites. However, on 23rd January 1901 Marconi succeeded in sending and receiving messages between The Lizard station and the Isle of Wight 300 km (186 miles) away using 46 m (150 ft) masts.[3] This was a world record for long-range wireless propagation, and was not just over the horizon but was some four times the optical range. Using the formula above the expected range over water would be about 75 km (47 miles) taking an average value for aerial height. Marconi was beginning to be proved correct in believing in his dream of long-range communication, and the scientists had to put on their thinking caps to explain the phenomenon. This event between the Isle of Wight and The Lizard on 23rd January 1901 was described at its 30th anniversary as *Marconi's First Great Miracle*.[6] Copies of telegrams and a letter which record the long-distance event are shown in Figures 10, 11 and 12.[4]

Marconi even had a vision of achieving inter-continental wireless communication, and he was at this time taking active steps to be able to demonstrate trans-Atlantic operation using new large stations at Poldhu in Cornwall and Cape Cod in America.[3]

Further details of The Lizard station

It is apparent that by 1909 the operating equipment at The Lizard station had been moved to the larger hut, and the smaller hut was then used as a store,[10] and this is confirmed by the photograph dated 1911 in Plate 22 which shows an aerial feeder entering this larger building." The layout of the buildings at that time is shown in Figure 13.[10] The station in these later years is described as having a 25 cm induction coil with spark-gap transmitter and condenser made up of twelve Leyden jars and powered by lead-acid accumulators charged by a dynamo which was belt-driven by a 2-stroke petrol engine. A magnetic detector had replaced the coherer by this time. The frequency of operation was 500 kHz and the range was out to 80 km (50 miles) and increased later to 160 km (100 miles); there was also a provision for 1 MHz operation.[7] See Footnote 1 for information on frequency.

A local farmer, Mr T S Hendy, kept a diary of village events, and as he became involved in assisting the station there is considerable detail available on events at the station over the years 1901 to 1916. For instance, he assisted on tasks such as raising the gaff on the mast, fixing the aerial and fixing a gate. On many occasions using his horse drawn vehicles he met personnel from the local railway station at Helston and recorded the names of all those conveyed and he also delivered materials and equipment. The price for carriage between Helston and Housel Bay was 60p, or 90p if a wagon and pair were required. In 1910 the price for collecting and delivering items from the "motor house" in the village was 10p. As the nearby Poldhu station became busy the two stations worked closely together and the basic cost of carriage between Housel Bay and Poldhu was 35p. Items carried included timber, stone, sulphuric acid (from 1902), goods, cases and receivers. Personnel carried in date order were Messrs Kimpt, Linsey, Hobbs, Hughes, Woodward, Long, Dagleigh, Shepherd, Ballon, Murphy, Marian, Edwards, Owen, Goldsmith, Furniss, McCrea, Cross, Epworth, Prara, Bullock, Blinkthorne, Beillocke and Davis; the spellings here are as in the diary.[12]

Relationships with Lloyd's

In an earlier chapter it was seen that Lloyd's was keen to achieve a monopoly on ship to shore signalling, and indeed in 1883 had taken-over operation of the former The Lizard Signal Station. It was therefore of some concern to Lloyd's in 1901 when the Marconi company setup a number of coastal stations for the very purpose of ship to shore signalling by wireless telegraphy, so much so that discussions were held and an agreement reached between the two companies that Lloyd's would not install in their stations the wireless equipment of Marconi's competitors provided Lloyd's was allowed to retain its monopoly operations in signalling. The existing Marconi stations were either to be taken-over by Lloyd's, or disbanded if Lloyd's already had better sites.

In the case of The Lizard with the adjacent but separate Lloyd's and Marconi stations it was agreed in October 1901 that the Marconi station would escape immediate Lloyd's take-over whilst the Marconi company completed some wireless development trials which were expected to take about ten days. These trials actually took much longer and by January 1902 it was agreed that Marconi would keep control for a further time with the Marconi company passing messages to the adjacent Lloyd's site, though by mid 1902 Lloyd's was expressing impatience on the take-over and by the end of 1902 suggested that the Marconi equipment be moved into the Lloyd's building and the mast re-sited close to that building. Discussions and studies on this move continued throughout 1903 and 1904 but the problems were always made too great and the two stations continued operating independently with the wireless station still under Marconi control, though of course providing messages to Lloyd's.

By early 1905 it was accepted that the equipment should stay in the Marconi hut, but Lloyd's was now trying to assert authority by imposing its staff. This in turn led to frustration and friction because the Lloyd's men had not been trained in wireless telegraphy operation, and standards fell with ships even complaining at times of the poor sending standard and the Marconi company said it would be years before Lloyd's would be capable of taking-over. In 1906 Lloyd's stated it would be prepared to take legal action if the Marconi equipment were not moved to the Lloyd's building, but neither the move nor the legal action took place and the Marconi station continued to operate separately.[15]

A summary of the station and the present position

The growth of mercantile marine application of wireless telegraphy progressed steadily, and by the end of 1904 the number of ships carrying Marconi equipment had risen to 124.[3]

The Lizard station was busy in its first years helping in developing long-range operation, in assisting the Poldhu high power station and in proving multiple-station operation in the vicinity of that high power station. It became ever more busy with ship to shore wireless telegraphy traffic, providing for the first time from 1901 day and night all-weather communication with ships.

The station continued until 1908 with its call sign LD as a Marconi station at which time the Post Office took-over responsibility and changed the call sign to GLD. Marconi staff provided assistance until 1911.57 In 1910 the first distress call by a Post Office British coastal station was received when the ship *Minnehaha* signalled that she was in trouble off the Isles of Scilly. The Lizard station answered the call and passed information to Falmouth from where tugs were dispatched which succeeded in pulling the ship from the shore.[13] The station was closed in 1913 when a newly developed station was opened at St Just, which was named Land's End Radio and took-over, the GLD call sign.[5,7] The Lizard station was reopened in the 1914-1918 period under Admiralty control and was used to listen for enemy transmissions and to provide communication with airships which operated anti-submarine long-range sea patrols from a station about 8 km (5 miles) inland.[14] The Marconi station was closed finally in 1920 and the mast and equipment were then dismantled.[5]

The buildings were left in place when the equipment was removed, and these were at a later date joined together to provide living accommodation, the owner adopting the name "Marconi" for the house. See Plates 23 and 24. The original huts are still there, having survived the gales of a hundred years. The concrete base of the mast is also there measuring 1.1m x 1.1 m x 1.3 m high (45 x 45 x 50 inches) with a central hole of 46 cm (18 inch) diameter where the mast stood, see Plate 25.

The Marconi Company in 1953 placed a commemorative plaque in the hedge close to the buildings with the words in italics at the head of this chapter.[5] This is shown in Plate 26.

The huts are probably the oldest surviving purpose-built wireless buildings, and have been acquired by the National Trust who in 2000 restored their layout into the original form. The original operating hut has been provided with replica wireless telegraphy equipment to the layout as in 1903, and is open as a visitor centre on many days each year. More modern equipment has been installed in a separate room which may be hired for used by radio amateur operators. A photograph of the restored buildings is shown in Plate 27.

Footnote 1

Electro-magnetic radiation is characterised by a term "frequency", which is the rate at which its constituent electric and magnetic components oscillate. The velocity of propagation of the radiation is 3×10^8 metres per second, and this leads to the term "wavelength" which is the distance apart in the propagation direction where the components at any instant are equal. The three quantities are related by the following formula

$$\text{frequency} \times \text{wavelength} = 3 \times 10^8$$

where frequency is in Hertz and the wavelength in metres. Thus frequencies of 500 kHz and 1MHz correspond to wavelengths of 600 m and 300 m respectively.

Footnote 2

In the years after 1945 Soviet Russia made a claim that Aleksander Popov from their country was the real inventor of wireless in the 1890s. A detailed investigation into this claim was made by Charles Susskind of the University of California, and he in 1962 published a paper that declared that whilst Aleksander Popov was an able scientist he made no significant contribution to wireless development and was certainly not its inventor, and this is the presently accepted position.

References

1. The Early History of Radio - from Faraday to Marconi. G R M Garratt. Published by the Institution of Electrical Engineers in association with the Science Museum. 1995.
2. Hertz and the Maxwellians. J G O'Hara and W Pricha. Published by Peter Peregrinus Ltd in association with the Science Museum. 1987.
3. A History of the Marconi Company. W J Baker. Methuen and Company Limited. 1984.
4. Copies of photographs, telegrams and a letter of the Marconi Company provided by Roy Rodwell and Louise Weymouth.
5. "Notes on the Lizard Station" provided by the Marconi Museum, Chelmsford.
6. "Picture of Marconi" speech 1931, provided by the Marconi Museum, Chelmsford.
7. Landsend Radio. J Ainslie. Wireless World. September 1938.
8. The Lizard Wireless Station - an archaeological survey. Cornwall Archaeological Unit. 1998.
9. Radio Engineer's Handbook. F E Terman. McGraw Hill. 1950.
10. Engineer's report on the Lizard Wireless Telegraphy Station, 24/9/1909. Made available by British Telecommunications.
11. The Yearbook of The Lizard Women's Institute. 1950.
12. The diaries of Mr T S Hendy. 1874-1930. By kind permission of Mrs E Hendy.
13. Communications from William Hocking of Gulval.
14. U-Boat Hunters - Cornwall's Air War 1916-19. P London. Dyllansow Truran. 1999.
15. Correspondence 1901-13 between Lloyd's and Marconi International Marine Communication Company Ltd. held in 11 volumes under MS 31665 at the Guildhall Library, London.
16. Data from the Marconi Archive at the Bodleian Library.

Although not directly referenced in the text, the following list of books provide further interesting reading on early wireless.

a. The Early British Radio Industry. R F Pocock. Manchester University Press. 1988.
b. Syntony and Spark - The Origins of Radio. H G J Aitken. Princeton University Press. 1985.
c. Marconi, A Biography. W P Jolly. Constable and Co. 1972.
d. Guglielmo Marconi. The Marconi Company Ltd. 1984.
e. The Wireless Man. F A Collins. 1912. Republished by Lindsay Publications Inc. 1993.
f. The How and Why of Radio Apparatus. H W Secore. 1922. Republished by Lindsay Publications Inc. 1993.
g. Wireless. Its Principles and Practice. R W Hutchinson. University Tutorial Press. 1933.
h. Henley's 222 radio circuit designs. J Anderson, A Mills and E Lewis. Originally published 1924 and republished in 1989 by Lindsay Publications Inc.
i. Early Radio - in Marconi's Footsteps. P R Jensen. Kangaroo Press. 1994.

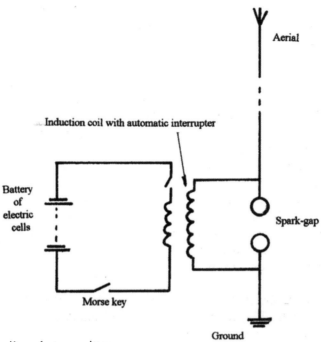

Figure 6.   Marconi's early transmitter

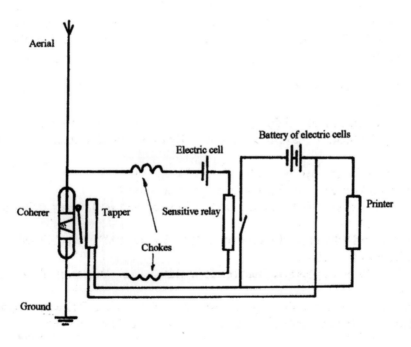

Figure 7.   Marconi's early receiver

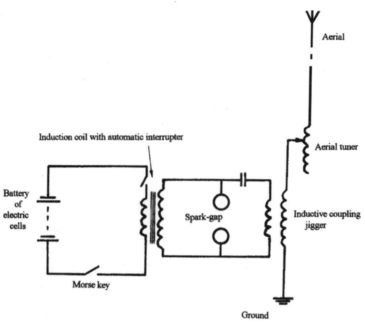

Figure 8.   Marconi's wireless transmitter with jigger

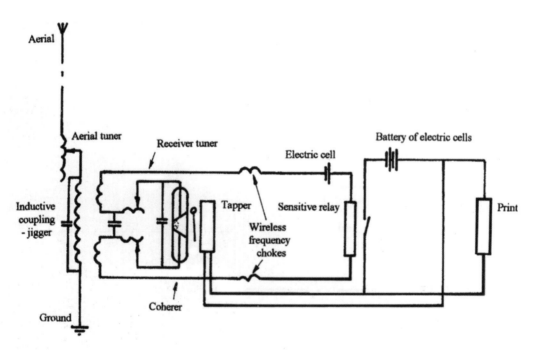

Figure 9.   Marconi's wireless receiver with jigger

40

Plate 8.	The Lighthouse from the west side in 1998
		Note the drawn curtain on the light showing this was before automation

Plate 9.	A view of the Lighthouse from the east

Plate 10.   Lloyd's Signal Station in 1998 as a residence

Plate 11.   The night boxes as residences in front of the signal station

View from a painting by Mr T Hart - By kind permission of Mr D M Todd

Plate 12.   The submarine cable brought ashore with the help of local fishermen

Plate 13.   Track of the submarine cable up the cliff

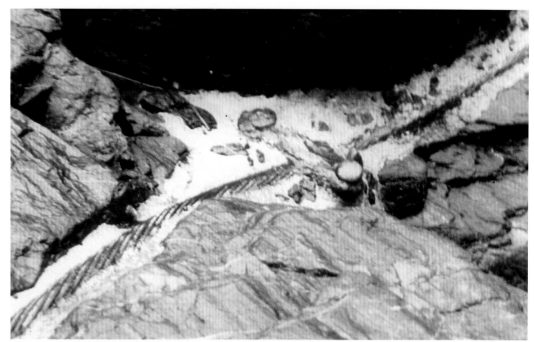

Plate 14.   A cleated section of the cable

Plate 15.   A damaged section of cable showing the inner conductor and armouring

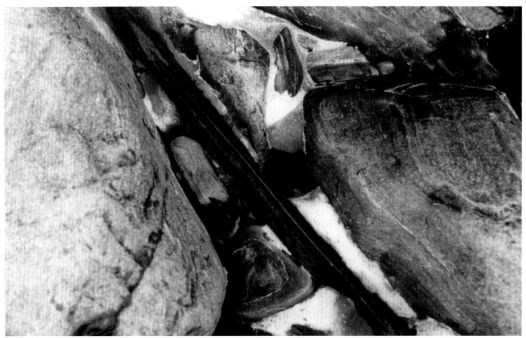

Plate 16.   The cable exposed on the beach in 1996. Note the gutta percha insulation

Plate 17.   The compressed-air foghorn in 1998

Plate 18.   The electric foghorn (in front of the railings)

Photograph by kind permission of Marconi plc

Plate 19.   The Lizard Wireless Telegraphy Station in about 1903

46

Photograph by kind permission of Marconi plc

Plate 20.   The Marconi station in about 1903

Photograph by kind permission of Marconi plc

Plate 21.   The interior of the station in about 1903

47

Photograph by kind permission of Marconi plc

Plate 22.   The wireless telegraphy station in 1911

Plate 23.   The Marconi station in 1998 as a residence

POST OFFICE TELEGRAPHS.

ENNISCORTHY

6.30 P.M.

Jan. 23, 1901.

Handed in at the Lizard Village Office at 5.32 p.m.
Received 6.24 p.m.

TO DAVIS KILLABEG ENNISCORTHY IRELAND

COMPLETELY SUCCESSFUL KEEP INFORMATION PRIVATE WILLIAM.

By kind permission of Marconi plc

Figure 10. Telegram from The Lizard on 23rd January 1901 announcing success of the "over-the-horizon wireless telegraphy trial

POST OFFICE TELEGRAPHS.

ENNISCORTHY

8.45 P.M.

Jan. 23, 1901.

Handed in at the THREADNEEDLE ST. Office at 6.35 p.m.
Received here 7.0 p.m.

TO JAMESON DAVIS KILLABEG ECORTHY.

Confidential Marconi completely successful ordinary apparatus Lizard to St. Catherines one nine six miles please keep this absolutely secret for present.

FLOOD PAGE.

Figure 11. A further telegram on the same day

MARCONI'S WIRELESS TELEGRAPH
COMPANY, LTD

TELEGRAMS,
EXPANSE, LONDON.

A.B.C. CODE USED.
TELEPHONE Nº 2748 AVENUE.

18, Finch Lane,

London, E.C.

11th February, 1901.

Dear Sir (or Madam),

I am instructed to inform you that
Mr. Marconi has established communication
between St. Catherine's (Isle of Wight)
and the Lizard (Cornwall) a distance of
196 miles. It is a matter of congratula-
tion that this long distance has been
successfully accomplished.

I am,

Yours faithfully,

HENRY W. ALLEN,

Assistant Manager & Secretary.

By kind permission of Marconi plc

Figure 12.   The Marconi Company announces to the world in February 1901 the achievement of
over-the-horizon wireless communication
(The distance of 196 miles in Figs 1 & 2 is stated by Marconi plc to be in error and
should read 186 miles)

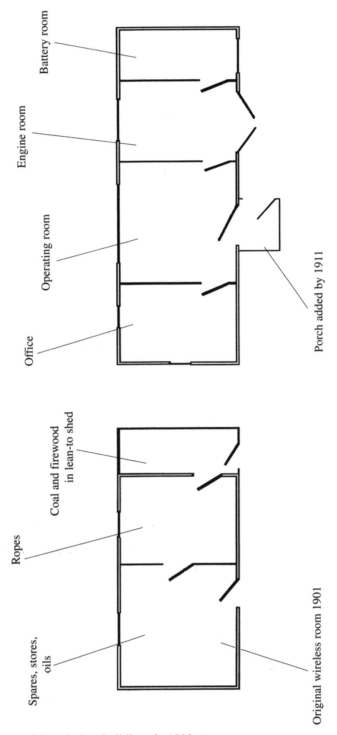

Figure 13.   Layout of the wireless buildings in 1909

Undersea Bell

The Lizard bell

On the western side of the sandy beach at the centre of Housel Bay there is a course of stonework on the side of the cliff, as shown in Plates 28 and 29, and many people ponder its purpose. During the winter, when the sea tides have moved much of the sand away from the beach, a length of armoured cable may be seen protruding from below this stonework as shown in Plate 30.

In the 10 year period (1868 to 1877), prior to the installation of the foghorn, the total tonnage of ships, which were wrecked at The Lizard during fog, amounted to nearly 4500 tons, with the average loss being 640 tons. In the 10 years after the foghorn was installed (1878 to 1887) the total of losses increased to 8100 tons with the average loss being 2000 tons. The years 1888 to 1897 saw a decrease to 5500 tons, but the next ten years (1898 to 1907) saw a very large total increase to 20500 tons with an average loss of 3400 tons.[1] The effectiveness of the foghorn was, therefore, questionable, and no new system was forthcoming.

Trinity House, in 1910, in an attempt to overcome this problem installed an undersea bell nearly 3.2 km (2 miles) to the south of The Lizard Head on the sea-bottom at a depth of 58 m (161 ft), and this was connected electrically to a submarine cable which came ashore at Housel Cove and was then taken underground to the lighthouse. This is the cable a part of which can still be seen at the western side of the cove. During fog the bell was rung electrically from the lighthouse, the aim being to get ships to steer near the sound of the bell and thus away from the coast. The bell and its associated parts is reported as weighing 380 kg (7½ cwt) and was mounted in a tripod shaped frame, the overall assembly weighing 2290 kg (2¼ tons).[2]

No records have been found regarding the effectiveness of the bell, but the violence of the sea caused it to be overturned on several occasions, and this and the unreliability of the equipment caused it to be withdrawn from service in 1914. The bell had been rung on 207 days.[2]

No new effective warning system in fog was forthcoming, and the toll of shipping lost at The Lizard continued, reaching a peak in the 1930s.[1] From there on the losses decreased dramatically, as ships became equipped with radar and electronic navigation aids.

The first hint that this cable on the beach might have been associated with the bell came from Mrs Hart of The Lizard. The information on its installation and operation is entirely dependent on letters received from Trinity House and two of these are shown in Figures 14 and 15.

Other bells

Trinity House had used electrically operated submarine bells from 1907, though the usual installation was with the bell suspended beneath a light vessel and rung from there. The results of trials led to a standard choice of a 100 kg (220 lb) bell which rang at a frequency of 1215 Hz.[3]

An electrically operated submarine bell resting on the sea bottom and driven from the shore was first installed for Trinity House by The Submarine Signalling Corporation at the South Stack Lighthouse, Holyhead in 1909 and in the following year at The Lizard. The bell for a typical sea-bed installation at that time is described in Reference 3 as mounted in a steel tripod of 7.6 m (25 ft) height and 6.4 m (21 ft) spread and with a weight, including bell and heavy cast-iron feet, of about 3400 kg (3.3 tons). A lifting chain was provided with its end buoyed. The bell together with its hammer and electrical mechanism hanging from the top of the tripod weighed about 340 kg (750 lb). The submarine cable to the shore had two power conductors for operating the hammer to ring the bell and two signal conductors to monitor the hammer position. Dynamos in the shore station provided the necessary electric current of about 3.5 amps and rotary time switches also at the station connected the power alternately to the bell and to loads to provide the necessary ringing intervals.[3]

There were 32 electrically operated undersea bells around the shores of Britain and Ireland in 1912 of which two were shore driven with most of the others hung from light vessels.[3]

By 1912 two hundred ships had been fitted with the necessary receiving equipment which consisted of a pair of microphones mounted in small water-filled tanks, inside the hull on each side of the bows, below the water line. The microphones were connected to earphones, and the navigation officer by comparing the loudness of the bell from the microphones on each side of the bows could obtain an estimate of the direction of the bell. Ships having the equipment included the *Olympic, Mauretania* and *Suevic*.[3]

By 1923 the Submarine Signalling Corporation had installed 127 electrically operated bell sending stations around the world, mainly on lightships, and 30 of these were around the shores of Britain and Ireland. By this time there were no bells mounted on the sea bottom,[4] and we may assume that the sea forces were too severe to achieve a reliable operation from the shore of a bottom-mounted bell.

References

1. The figures on shipping losses quoted here have been derived from shipwreck records kept by Mr A E (Bert) Jane, and are included by kind permission of Mr A A Jane.
2. Letters from Trinity House. 1999. These are included here as Figures 14 and 15.
3. Submarine Signalling. The Submarine Signal Company, London. 1912.
4. Submarine Signalling. The Submarine Signalling Corporation, London. 1923.

Trinity House Lighthouse Service

Serving the Mariner

Courtney Rowe
Trenook
Church Cove
The Lizard
Helston
Cornwall
TR12 7PQ

Our ref: 55698

18 January 1999

Dear Mr Rowe

Submarine Bell at Lizard Lighthouse

Thank you for your letter dated 9 January regarding the above.

I can advise you that the bell was laid 1 mile and 3 cables due south from the lighthouse in 161.5 feet of water at Low Water Spring Tide. The cable you mention connected the bell to dynamos in the engine room of the lighthouse.

The bell was laid on 19 May 1910. It was withdrawn in June 1914. The Elder Brethren of Trinity House thought that if regularly maintained it would be a valuable aid to navigation. However the uneven seabed and strong tides meant that the cost of maintenance was unjustifiable. The bell was in action for 207 days before it was discontinued. Problems were caused by the tripod the bell was mounted on overturning several times and also defects with the bell. The bell failed on:

2 July 1910
4 July 1911
31 July 1911
27 December 1912

The tripod the bell was mounted on was 21 feet high with a spread of 12 feet radius. The entire structure weighed about 2 ¼ tons, the bell accounting for 7 ½ cwt. The total cost was £2,405.

I hope this information is of use to you.

Yours sincerely

Karl Ree

Karl Ree
Human Resources & Communication

Administration Directorate Trinity House Tower Hill London EC3N 4DH
Tel: 0171 480 6601   Fax: 0171 480 7662

By kind permission of Trinity House

Figure 14.   Details of the undersea bell

Trinity House Lighthouse Service

Serving the Mariner

Courtney Rowe
Trenook
Church Cove
The Lizard
Helston
Cornwall
TR12 7PQ

Our ref: 56655

11 February 1999

Dear Mr Rowe

Submarine Bell at Lizard Lighthouse

In response to your letter dated 8 February I can advise:

1.   The bell would be heard any where in the ship that was below the water line. With listening equipment a bearing could be taken from the bell. Sound travels three times further underwater. Submarine bells could have a range of up to 15 miles.

2.   I cannot find any evidence of a submarine bell being used in conjunction with a fog horn to indicate distance from the shore. I do not know the modus operandi of the bell but it was either sounding continuously when in commission or used as a fog signal.

3.   I do not know the cross section of the cable.

4.   I do not know how often the bell was sounded or what power was required.

5.   We have no drawings or photographs of the bell and tripod.

Sorry I could not be of much further help but Trinity House lost the majority of its records in the Blitz.

Yours sincerely

Karl Ree

Karl Ree
Human Resources & Communication

Administration Directorate Trinity House Tower Hill London EC3N 4DH
Telephone: 0171 480 6601   Fax: 0171 480 7662
website at http://www.trinityhouse.co.uk

By kind permission of Trinity House

Figure 15.   Further details of the undersea bell

Plate 25.   The Marconi buildings from the rear in 1998

Plate 26.   The base of the mast in 1998

GUGLIELMO MARCONI
whose pioneer work in
wireless telegraphy
for the safety of all seafarers
was furthered in this building
during the first years of
the century

PLAQUE PRESENTED BY
MARCONI'S WIRELESS TELEGRAPH COMPANY LIMITED
CHELMSFORD ESSEX 1955

Plate 22.   The commemorative plaque

Plate 27.   The station in 2000 after restoration by the National Trust

Plate 28.   Stonework at the west side of Housel Cove

59

Plate 29.
A close-up of the
stonework in summer

Plate 30.   The armoured cable showing below the stonework in winter

Plate 31.   Derelict radar building in 1999

Plate 32.   Base mounting for a former radar mast

61

Plate 33.   The look-out post once used by the Coastguard and now used by the National
              Coastwatch Institution

Plate 34.   Royal Observer Corps - Orlit post in 1999

Plate 35.   Royal Observer Corps - underground post in 1998

Plate 36.   The mast, aerial and transmitter cabin for the radio beacon to the
north of the lighthouse

Plate 37.   Cabins and masts of the navigation station

Plate 38.   Receiving aerials of the Differential Global Positioning System on the
west tower of the lighthouse

RAF Pen Olver

A radar station was first set up in the area local to The Lizard at Dry Tree on Goonhilly Downs some 10 km (6 miles) inland. This was a Chain Home (CH) station with powerful transmitters, high masts and large aerials for detecting aircraft at long range and plotting their tracks; it was operational by June 1940. The equipment was very effective against high-flying aircraft, but because of its radio frequency of 27 MHz with the corresponding wavelength of about 11 metres, the equipment was not suitable for detecting low level aircraft at sufficiently long range. For this reason the Dry Tree station was equipped in Autumn 1940 with another radar which used a higher radio frequency; this was known as a Chain Home Low (CHL) radar and had been designed to give improved performance at low level. This latter CHL radar however did not perform well at Dry Tree and the low-level function was moved to a new station on the coast, at Housel Bay, the station taking the name RAF Pen Olver from a headland in the bay.[1]

RAF Pen Olver was used for the detection of low flying aircraft and also shipping, and was located in buildings, including several Nissen huts, near Lloyd's Signal Station. Personnel were also billeted at the Housel Bay Hotel and in the village, and the former Marconi huts were used as an officer's mess.[2] A wooden hut now in a garden in the Church Cove area was once a billet for personnel and is now used as a workshop. Some of the social aspects of life of the station personnel may be read in Reference 2, below.

RAF Pen Olver became operational in April 1941 as a Chain Home Low station. The CHL radar installed then operated at a radio frequency of about 200 MHz corresponding to a wavelength of 1.5 metres, and with this relatively short wavelength the aerials could be made sufficiently small to be mounted on top of huts and could be turned to scan in azimuth. The first CHL radars operated with two such huts, one for the transmitter and the other for the receiver and were separated by about 200 m (650 ft) and this was probably the first type of radar to be installed at the station. Later it was possible to combine the transmitter and receiver functions into one set of equipment and a combined aerial on a single hut,[3] and this type of equipment with a single aerial was installed in the building known as the "night box" a short distance in front of Lloyd's Signal Station;[2] this building having been built by the Signal Station in a location where it was shielded from the powerful beam of the lighthouse to make signalling by lamp easier at night. The lattice type scanning aerial of the radar was mounted on top of the "night box" and measured about 8 by 3.5 metre (26 x 11 ft).[4] Information on detected aircraft provided by the radar was transmitted by landline to Dry Tree CH station where the data was passed via the filter room to fighter control.[3]

In 1942 an additional radar, designated Type 31, was installed at RAF Pen Olver. This used a higher radio frequency of 3000 MHz (10 cm wavelength) enabling improved low level detection to be achieved with ships plotted to a high degree of accuracy. It was known as a Chain Home Extra Low (CHEL) radar and with its low power was housed in a transportable wooden cabin, known as a Gibson Box, measuring 3.3 m by 2.1 m and 2.4 m high (11 x 7 x 8 ft). It was surmounted by a hand-turned paraboloid aerial.[3]

During 1944 the site was provided with a further CHEL radar known as Type 56. This again used a 3000 MHz frequency but had a high power transmitter. The aerial, a paraboloid of 3 m (10 ft) diameter, was in this case mounted on a turntable at the top of a 56 m (184 ft) high wooden mast to the rear of Lloyd's Signal Station, with the equipment being located in a hut. The aerial could be scanned in azimuth at speeds up to 6 revolutions per minute or could be pointed in specific directions in azimuth as required, and with its improved equipment, higher power and its larger and higher aerial this radar gave a very significant improvement in capability.[3]

All these radars at RAF Pen Olver provided coverage for detection of low flying aircraft, but in addition they had a capability at higher elevations and thus were also able to provide data at these levels to the Dry Tree station. The Pen Olver radars also had an important role in plotting the progress of allied shipping convoys and in warning of attack by gun boats, ships or submarines,[3] and there was a naval personnel presence at the station to report into headquarters at Falmouth any enemy ship activity and especially E-boat presence. Barbed wire, armed guards and two anti-aircraft guns protected the station.[2]

A map provided by the Royal Air Force Museum at Hendon shows that there were considerations at one time to obtain a new site for the radar on the west side of Housel Bay, but no construction was undertaken.

In March 1945 RAF Pen Olver was still operating the CHL and the CHEL Type 31 and Type 56 radars but by October 1945 just the CHL and Type 56 CHEL radars were in use.[3] Today all that can be seen of the RAF station are the remains of brick buildings to the rear of Lloyd's Signal Station and the concrete base points of a mast, as shown in Plates 31 and 32. It is considered that these remains are from a larger radar and aerial installed after 1945 with the aerial scanning on a low gantry over the building but no details are available. The night box has now been converted into living accommodation.

The station was subjected to an aerial attack during its operational period but all the bombs landed in nearby fields and did no harm to the station.[5]

References

1. Radar Installations in SW Cornwall during the Second World War. Ian Brown. The Historical Radar Archive, Peebles, Scotland. Paper presented at a Trevithick Trust Symposium on "Communications in Cornwall". June 1997.
2. Cornish War and Peace. V Acton and D Carter. Landfall Publications. 1995.
3. "RAF Pen Olver". Private paper provided by Ian Brown, The Historical Radar Archive. 1997.
4. Fox's Signal Station, Bass Point, The Lizard. S B Smith. 1992. Paper to The National Trust.
5. Communication with H Stevens, The Lizard.

Other Installations at the Bay

The Coastguard

The Coastguards had a presence at The Lizard from the nineteenth century,[1] and in 1951 Lloyd's passed to them the responsibility for reporting passing ships and for signalling with them. The Coastguards then occupied Lloyd's Signal Station.[2] The Lizard Women's Institute Year Book of 1965 records that some 22,000 ships passed in 1964 and almost 20,000 in 1965.3 Eventually the Coastguards left this station and moved to a purpose-built lookout post at Bass Point, (see Plate 33) With the passage of time most boats and ships became equipped with radio transmitters and receivers and by 1992 it was judged that this Coastguard Station was no longer necessary, and it was then closed.[1]

In 1995 a voluntary body, the National Coastwatch Institution, reopened the site as a part-time lookout for dangers at sea and on the cliffs, and this body still operates there.[4] Further information on the overall aspects of the Coastguards and Coastwatch at The Lizard may be found in References 1 and 4.

Royal Observer Corps

A short distance inland on Lloyd's Road is a site which maps now show to be a reservoir. In fact it was a Royal Observer Corps (ROC) site which dates from 1941, when the original, probably timber, building was built. The buildings from that period have gone, though there are presently the remains there of two subsequent buildings neither of which are now used. One of these buildings was an "Orlit Type A" above-ground observation post which dates from 1951 and was used for plotting aircraft tracks in the "atomic" Cold-War period and reporting them to a central station. The standard design for this type of building measures 3.05 x 2.03 metres (10 x 7 ft) with part of the building roofed for shelter and storage and with an adjacent open-top observation section, see Plate 34.[5,8]

Alongside this "Orlit" post is an underground monitoring post built in 1958, and this was used in the "nuclear" Cold-War period. All that is visible above ground, see Plate 35, are a mound of earth, access hatch, an air ventilator and sensor arms. This post was designed for monitoring nuclear explosions and fall-out. The hatch when opened reveals a steel ladder which gave the three operators access to a room 4.6 m (15 ft) below measuring 4.6 x 2.3 m (15 x 7.5 ft). A semi-rotary hand pump is mounted on the wall, and a small room houses a chemical closet. A bomb power indicator probe and a survey meter probe, both above surface, were connected to

The site closed in 1991.[8]

instruments in the room below.[5,8]

Radio Beacon

In 1950 Trinity House installed a radio calibrator at the lighthouse and this operated in daylight hours with the signal LD. This was changed in 1954 to a radio beacon, which transmitted the letters LZ at a frequency of 298.8 kHz in Morse code in a defined manner and had a reception range of up to 130 km (70 nautical miles). A ship by monitoring the signal from the beacon with a direction-finding radio could determine its direction, and by additionally monitoring transmissions from other beacons at different known locations the ship could determine its position.[6]

A mast held one end of a 21 m (70 ft) long "T" shaped transmitting aerial, the other end of the aerial being supported by the lighthouse tower, see Plate 36, and the centre point connected to an operating cabin below containing the transmitter. The sequence of transmission from the beacon was four signals of LZ lasting 19.6 seconds, followed by a dash of 25 seconds and then two more LZ signals lasting 9.6 seconds, this sequence thus taking 54.2 seconds; the whole sequence was repeated every 6 minutes. This beacon operated until early 1999.[6]

Naval navigation station

Close to the location of the former RAF Pen Olver station were a mast and unmanned cabins which formed a segment of a RN navigation system known as Hyper-Fix, see Plate 37. One cabin contained the operating electronics equipment, the other a standby generator.[7] The station closed in the year 2000.

Differential Global Positioning System

The international navigation service known as the Global Positioning System (GPS) uses 24 satellites that are circling the earth in accurately known orbits. A radio receiver on earth, whether in a ship, aircraft or elsewhere can determine its position within an accuracy of plus/minus 100 metres (330 ft) at 95% probability by monitoring the radio transmissions from four of the satellites. This is a world wide system that operates by day and night and in all weathers.[6]

Trinity House at The Lizard Lighthouse introduced an additional facility known as Differential Global Positioning System (DGPS) in 1998. Aerials for reception of GPS signals from the satellites are mounted on top of the west tower of the lighthouse see Plate 38, and as the position of this tower is known very accurately the errors in the GPS system at this location can be determined. These errors are calculated and then transmitted, enabling ships equipped with GPS and DGPS receivers and out to 90 km (50 nautical miles) from the coast to determine their positions to an improved accuracy of plus/minus 10 metres (33 ft) at 95% probability. The aerial of the former radio beacon (Plate 36) is used to transmit the error data at a radio frequency of 284 kHz. The service is suitable for use by vessels ranging from liners and cargo ships to yachts and fishing boats, and may be used for navigation in moving ships and for accurate positioning when stationary.[6] There are a total of 12 DGPS stations around the coast of Britain and Ireland, each with its own transmission frequency, which give overlapping coverage.[6]

References

1. The Lizard in Landewednack - A Village Story. The Lizard History Society. 1997. Section by N Green from page 76.
2. Cornish Lights and Shipwrecks. Cyril Noall. 1968.
3. The Yearbook of The Lizard Women's Institute. 1965.
4. The Lizard in Landewednack - A Village Story. Section by Lynn Briggs from page 78.
5. Twentieth Century Defences in Britain - an Introductory Guide. Editor B Lowry. Council for British Archaeology. 1995.
6. Papers and comments from Trinity House.
7. Discussions with service engineer.
8. Data sheet by Mr. Lawrence Holmes, The Royal Observer Corps Association

Appendix

Some other Cornish wireless/radio sites

Poldhu

The following words are on the east side of the Marconi column at Poldhu. *"One hundred yards north east of this column stood from 1900 to 1933 the famous Poldhu wireless station. Designed by John Ambrose Fleming and erected by the Marconi company of London, from which were transmitted the first signals ever conveyed across the Atlantic by wireless telegraphy. The signals consisted of a repetition of the Morse letter "S" and were received at St John's Newfoundland by Guglielmo Marconi and his British associates on 12.12.1901".*

At the same time as The Lizard Wireless Telegraphy Station was being built, Marconi was constructing another and larger installation on a site 10 km (6 miles) away on the cliff top at Poldhu. This was to undertake tests to determine if wireless transmissions would cross the Atlantic Ocean and be received in America, and if the tests were successful, it was the intention to set-up a commercial trans-Atlantic wireless telegraphy service. Because of the distance involved everything was made on a large scale. For instance, the aerial system was supported by 20 masts each 61 m (200 ft) high and arranged in a circle 61 m (200 ft) in diameter and the transmitter was not powered by a battery, as at The Lizard, but by a 25 kW motor driven generator.[1,2] The short period of ever-larger wireless stations was just beginning.

By the Autumn of 1901 the station at Poldhu and a receiving station at Cape Cod, USA 5000km (3100 miles) away were ready to undertake tests, but severe gales at Poldhu in September demolished the aerial there, and a hurriedly built replacement was only 48 m (160 ft) high and supported by just two masts. But then the aerial at Cape Cod, to the same design as the first Poldhu aerial, was, in its turn, demolished in a gale in November.[1,2]

In the face of these two disasters it was decided to make a limited trial with the western receiving point at the nearest land at St John's in Newfoundland, Canada 3400 km (2100 miles) away. The aerial used at St John's was a wire some 150 m (500 ft) long held aloft in difficult conditions by a kite. Marconi listened using earphones over several days and on 12th December was sure he had heard, amongst the wireless static noise, the agreed signal - the letter S in Morse code transmitted from Poldhu, but the signal was only heard for a few instants and was not strong enough to work a printer. The only record was, therefore, in the minds of Messrs

Marconi and Kemp, who claimed they had heard faint signals, but this was not sufficient to convince the doubters. The frequency of the transmission was stated by Marconi to be 820 kHz (corresponding to a wavelength of 366m), though others have considered a figure of about 100 kHz (a wavelength of 3000m) more likely.'2 The achievement in December 1901 of trans-Atlantic wireless communication was described later as "Marconi's Second Great Miracle".[3]

Marconi knew that he had succeeded, and in February of 1902 using a wireless receiver on the SS *Philadelphia* he set sail from England across the Atlantic receiving and recording signals from Poldhu as he travelled west. The messages recorded on a printer were received out to 1100 km (700 miles) by day and to 2500 km (1550 miles) by night, and aural signals were heard at night out to 3400 km (2100 miles). This was enough evidence to convince everyone that trans-Atlantic operation was possible. The frequency of transmission in these latter tests was 182 kHz (1650 m).[1,2] The discoveries that such long-range propagation was indeed possible and that reception was better at night were to open new and challenging avenues for research, which were to occupy scientists for years.

Arguments with the undersea cable companies were to preclude Marconi setting up a station in Newfoundland, and he accepted an offer to be allowed to build a station at Glace Bay, Nova Scotia, Canada 4000km (2500 miles) away from Poldhu. This new station had four masts 61 m (200 ft) high supporting the aerial and the power input had been increased to 75 kW. Its signals were first heard at Poldhu on 28th November 1902, though it was to be the 5th December before readable messages were received there. A limited two-way wireless telegraphic service between Glace Bay and Poldhu was started, but it was not reliable and in March 1903 was brought to an end when the aerial at Glace Bay was brought down by ice. It was decided that plans for a wireless telegraphic service should await the development of even larger stations.[1,2]

By 1905 a new aerial had been built at Glace Bay and occupied a circle 700m (2200 ft) across. The frequency now chosen was 82 kHz (3660 m), and it had been realised that with the trend to ever lower frequencies and corresponding longer wavelengths the Poldhu site would not be large enough to accommodate the large aerials necessary to provide a reliable service. The possibility of using Goonhilly Downs was considered,[4] but in the event a new station was built at Clifden, Ireland, where the frequency was selected to be 45 kHz (6666 m); the power was 300 kW with voltages up to 20 kV. Then at last in October 1907 a trans-Atlantic wireless telegraph service was started between Clifden and Glace Bay, and by February 1908 had handled 120000 words.[1,2]

Meanwhile the Poldhu station took-on new tasks. In a commercial role it became until 1922 a wireless telegraphy station operating with stations in Europe and with ships out in the Atlantic Ocean, and it also conducted important wireless research until 1934 when the station closed.[1,2]

The following words are on the north side of the Marconi column. "*The Poldhu wireless station was used by the Marconi Company for the first trans-oceanic service of wireless telegraphy which was opened with a second Marconi station at Glace Bay in Canada in 1902. When the Poldhu station was erected in 1900, wireless was in its infancy, when it was demolished in 1933, wireless was established for communication on land, at sea, and in the air, for direction finding, broadcasting*

72

and for television".

Porthcurno

The Eastern Telegraph Company had by 1900 built-up a major site west of Penzance at Porthcurno for operating submarine telegraph cables to places such as Africa, India, The Middle East, the Far East, Australia, New Zealand, Spain, Gibraltar, Malta, France and also the Isles of Scilly and this activity was to continue for many years.[5]

The company, however, at the beginning of the twentieth century was aware of the progress being made in wireless telegraphy and that this might in time lead to competition with cable communication. It decided to set up a small research facility at Porthcurno in order to become acquainted with the capabilities of wireless, and this facility made its own wireless equipment and undertook experiments with it. Across Mounts Bay could be seen the Marconi aerial masts at Poldhu and at Housel Bay, and in 1902 the Eastern Telegraph Company erected its own 170 ft mast with aerials on the cliff-top near Porthcurno and assembled receiving and possibly transmitting equipment in an adjacent hut; this was used in particular for monitoring the Marconi transmissions and continuing research activity until 1914.[5]

A more full description of the wireless activity at Porthcurno may be found in Reference 5.

Land's End

The Lizard Wireless Telegraphy Station had served the south west approaches well in ship to shore wireless telegraphy in the first decade of the twentieth century, but eventually a new station was required to deal with the increasing traffic and to take advantage of new equipment developments. This was built at St Just under the name of Land's End Radio and came into service in 1913; it was given the call sign GLD of The Lizard station it replaced.[6]

This Land's End station, still operating in wireless telegraphy mode, used spark-gap transmitters at first though more powerful than at The Lizard. A 15 HP engine powered a DC generator, this being duplicated for reliability, and fed accumulators of 36kA capacity. The spark transmitter was fed with 5kW of power, giving an official daytime range of 400 km (250 miles) and by night the range was often out to 1600 km (1000 miles). A second spark transmitter was fed at 1.5 kW, and a third spark transmitter powered by a dry battery was kept for emergency purposes. The wireless frequencies were 500 kHz, 660 kHz and 1 MHz (wavelengths of 600, 450 and 300 metres). The two masts were 61m (200 ft tall).[6]

The station was modernised in 1932 by which time spark transmitters had been superseded by valve driven circuits. The power into the aerial by the new transmitter was 5kW at a frequency of 500 kHz (600 metres), and the range was 800 km (500 miles) by day and 2400 km (1500 miles) by night. This transmitter was now powered from the mains and a second 1 kW output transmitter powered by an alternator was available for emergency use.[6]

The transmitters described above operated in the wireless telegraphy mode but the station was by 1932 also equipped with a short-range radio telephone operated with a battery powered 300 W transmitter. This operated between 1.2 and 3 MHz and had a range of 160 km (100 miles). Two direction finders were also installed.[6]

By 1938 about 100 ships communicated with the station each day and it probably handled more traffic than any other coastal wireless station. The station was further updated at other times in its lifetime with traffic building up to 500 per day. In 1979 the receiving equipment was moved to Sennen to the site of the former RAF Sennen station which had operated Gee navigation equipment, and this move provided more space and quieter operation. The transmitters remained at St Just.[7]

The station once requiring three shifts, each with six telegraphists, was eventually unmanned and remotely controlled from Stonehaven in Scotland; it closed in 2000.[8] A more complete account of the station may be found in Reference 9.

Mile End

On the main road 1.6 km (1 mile) inland from The Lizard Village is a row of buildings at a place now known as Mile End. The houses here provided residence for the staff of a former wireless station and are now private homes. The masts and aerials were erected in the field behind.

This station formed part of Royal Naval Air Station, Mullion which from 1916 operated antisubmarine airships at up to 300 km (190 mile) from the coasts of Cornwall and Devon. The airships were based about 6 km (4 miles) inland at Bonython near Cury Cross Lanes (adjacent to the present windmill site) and at other sites in Cornwall. The new wireless station at Mile End operated in a direction-finding role with call signs BVY and BVW, and the airships when on patrol transmitted by wireless their call signs at regular intervals, and from the received signals the Mile End station was able to determine the bearing of each airship at these times. With the aid of another bearing from a second station at Prawle Point in Devon the tracks of the airships flights were plotted. A story of the operations of the airships, and aircraft which were introduced later, may be read in Reference 10. As mentioned earlier the Marconi station at Housel Bay was pressed back into service at this time to provide a wireless telegraphy communication service with the airships, and was also able to listen for enemy transmissions.[10,16]

The Year Book of Wireless Telegraphy 1922 describes the direction finding station as still under Admiralty operation with a location of 49 59 6.3 N and 5 12 24.1 W It had by then been reduced to a single call sign BVY at a frequency of 667 kHz (a wavelength of 450 metres).[11] In 1924 the station was transferred to the Post Office and then at a new frequency of 375 kHz provided a commercial direction finding service for all ships. Any ship requiring to know its bearing with respect to the station would transmit to Mile End its call sign in Morse code followed by a long dash which would enable the operator at Mile End to fix its bearing and this was then transmitted back to the ship. The station operated in this role until 1936 when it was closed and the dwellings sold.[7]

Bodmin

By 1924 there were plans to construct very large wireless stations which would communicate with the distant parts of the Empire such as South Africa and Australia. The calculations had shown that these stations would have to be larger than any before both in aerial size and transmitter power. Aerials as tall as 245 m (800 ft) would be necessary with power to the transmitters of 1 MW. These were truly immense figures and stations would be expensive to build and operate, but decisions were taken to proceed.[1]

However, the Poldhu station, now concentrating on research, had been investigating the use of higher frequencies (shorter wavelengths), with which it found that much longer ranges were possible than were expected. In 1924 the station succeeded, for instance, in communicating directly with Australia at a frequency of 9.3 MHz (32 m wavelength) and with only 12 kW of power and a relatively modest aerial. Bearing in mind the work that had gone into development of the proposed new very large stations with their long wavelengths, and the readiness of the plans for the equipment, it was a difficult decision on whether or not to disclose this new information. In the event it was released and the plans for the big stations were abandoned in favour of the higher frequency (short wavelength) stations.[1] The aerials for these new "Beam wireless" stations were 87 m (300 ft) tall and supported by 5 masts. One of these stations as part of the "Imperial Wireless Chain" was constructed near Bodmin and became operational in October 1926; it contained the transmitters to Canada and to South Africa. The receiving station for these countries was near Bridgwater. Stations outside Cornwall provided communications with other parts of the world.[12,13,14,15] The following words appear on the south side of the Marconi column at Poldhu. *"From the Marconi Company's Poldhu Station in 1923 and 1924, Charles Samuel Franklin, inventor of the Franklin beam aerial, directed his short wave wireless beam transmission to Guglielmo Marconi on his yacht "Elettra " cruising in the south Atlantic. The epoch making results of these experiments laid the foundation of modern high-speed radiotelegraphic communication to and from all quarters of the globe".*

Footnote

A representation of the electromagnetic spectrum is shown in Figure 16, and a summary of frequencies and wavelengths used in the early years of the twentieth century for long-wave and short-wave operation in Figure 17.

References

1. A History of the Marconi Company. W J Baker. Methuen and Co Ltd. 1984.
2. The Story of Poldhu. A Chapter of Marconi History.
3. "Picture of Marconi" speech 1931, provided by the Marconi Museum, Chelmsford.
4. Data provide by the Marconi Museum, Chelmsford.
5. The Spies at Wireless Point. J E Packer. Porthcurno Telegraph Museum. 1991.
6. Landsend Radio. Wireless World. September 1938.
7. Communications from William Hocking, a former telegraphist at Lands End Radio.
8. Discussions with David Lancarrow, another former telegraphist at the station.
9. Watchers of the Waves. B Faulker. G C Arnold Partners. 1996.
10. U-Boat Hunters - Cornwall's Air War 1916-19. P London. Dyllansow Truran. 1999.
11. Year Book of Wireless Telegraphy. 1922.
12. Communications from Derek Reynolds.
13. Imperial and International Wireless Communications 1909-1946. A Chapter of Marconi History-6.

14. Wireless Beam Stations. The Engineer October 29th 1926.
15. The Big Beam - The Tetney Beam Station. Radio Bygones. April/May 2000.
16. Airships in Peace and War. J. A. Sinclair. 1934

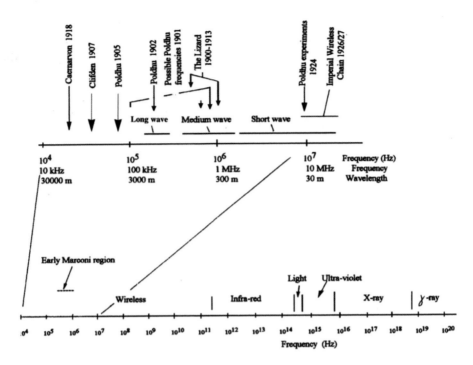

Figure 16.   The electromagnetic spectrum

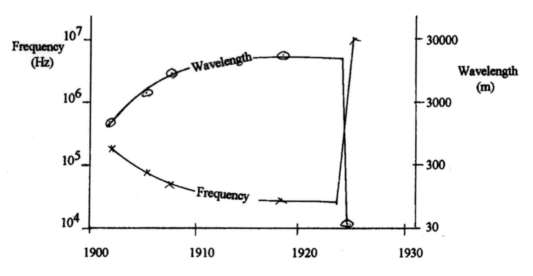

Figure 17.   Wireless frequencies and wavelengths used for long range
communications in the 20th century

Acknowledgements

The information in this book relates in the main to equipment installations which were in operation before the memory span of people still alive, and the author has relied heavily on printed information, company archives, photographs, interpretation of remains still on the ground and discussions with a number of people. He is grateful to all those who have helped in any way, and in particular for the help from the following people.

Mr Karl Ree of Trinity House for researching the history of the submarine bell and for making available the related reference documents. Also for providing information on Differential GPS and for arranging for engineers to check the sections in this book on the lighthouse, foghorn, undersea bell, Differential GPS and radio beacon and for providing corrections to these sections.

Mr Eddy Matthews, the keeper at the lighthouse, for providing information during guided tours.

Mr C L Fox of G C Fox and Company for making available for study the company archives from which most of the information in the chapter on The Lizard Signal Station has been derived, and for checking this chapter.

Mr Stuart Smith of the Trevithick Trust for providing a copy of his report to the National Trust on Lloyd's Signal Station and an engineer's report from BT on The Lizard Wireless Telegraphy Station. Also for making a number of constructive comments on the content of this book.

Mr A A Jane for permission to use information on shipwrecks compiled by his father Mr A E (Bert) Jane.

Mrs E Hendy for making available for study the diaries of Mr T S Hendy covering the years 1874 to 1930 which have been particularly relevant to the lighthouse and Marconi.

Mr David Todd for making available a copy of his report which provided most of the information on the cable to Spain, and for allowing a copy of his painting to be included.

The Lizard Women's Institute for making its yearbooks accessible and for agreeing that certain parts could be used. The photograph of the Marconi buildings in 1911 has been especially valuable and has also been used as a guide in the reconstruction of the buildings.

Mr William Hocking for providing information on Lands End and Mile End Radio stations and for checking the related sections and those on Marconi and Poldhu.

Mr Tony Pawlyn for researching the ship in the Marconi photograph.

Mr Roy Rodwell and Louise Weymouth of Marconi plc for making available photographs and documents and a copy of the book The History of the Marconi Company. Also Gordon Bussey, a historical consultant to the company for information.

Mr Eric Edmonds for making available his files on the history of communications in Cornwall.

Mr Ian Brown of the Historical Radar Archive for providing most of the information on RAF Pen Olver.

Mr Ron Julian for information on the Poldhu station.

Mary Godwin of Cable and Wireless for checking and providing comments on the section on Porthcurno.

Mrs Joan Hart for giving the first indication that the eastern cable might have been associated with a bell.

Captain Mike Tarrant, formerly of Trinity House for information on bells.

Mr David Lancarrow for giving a guided tour of Lands End Radio and for providing related information.

Mr Derek Reynolds for information on Bodmin Radio.

English Heritage and the Ministry of Defence for providing copies of photographs of buildings and aerial views and for permitting their use in this book.

The National Trust for continued liaison on the Marconi site.

The Guildhall Library, London for providing access to correspondence between Lloyd's and the Marconi company.

Charles Thurlow, Allen Buckley and Vernon Baldry of the Trevithick Society for providing valuable comments on the manuscript of this book.

Index

The letters P below refer to plates. Other numbers are page numbers.